Epstein-Barr Virus - New Trends

Edited by Emmanuel Drouet

Published in London, United Kingdom

IntechOpen

Supporting open minds since 2005

Epstein-Barr Virus - New Trends
http://dx.doi.org/10.5772/intechopen.87415
Edited by Emmanuel Drouet

Part of IntechOpen Book Series: Infectious Diseases, Volume 11
Book Series Editor: Shailendra K. Saxena

Contributors
Hugo Manuel Lopes de Sousa, Joana Patrícia Costa Ribeiro, Mafalda Basílio Timóteo, Andi Arus Victor, Sylvie Ranger-Rogez, Zheming Lu, Ying Liu, Hongying Huang, Pia Bernasconi, Paola Cavalcante, Federica Bortone, Renato Mantegazza, Letizia Scandiffio, Emmanuel Drouet

First published in London, United Kingdom, 2021 by IntechOpen
IntechOpen is the global imprint of INTECHOPEN LIMITED, registered in England and Wales, registration number: 11086078, 5 Princes Gate Court, London, SW7 2QJ, United Kingdom
Printed in Croatia

British Library Cataloguing-in-Publication Data
A catalogue record for this book is available from the British Library

Additional hard and PDF copies can be obtained from orders@intechopen.com

Epstein-Barr Virus - New Trends
Edited by Emmanuel Drouet
p. cm.
Print ISBN 978-1-83968-489-0
Online ISBN 978-1-83968-490-6
eBook (PDF) ISBN 978-1-83968-491-3
ISSN 2631-6188

For EU product safety concerns:
IN TECH d.o.o., Prolaz Marije Krucifikse Kozulić 3, 51000 Rijeka, Croatia, info@intechopen.com or visit our website at intechopen.com.

IntechOpen Book Series
Infectious Diseases
Volume 11

Emmanuel Drouet, PharmD, is a Professor of Virology at the Faculty of Pharmacy, the University Grenoble-Alpes, France. As a head scientist at the Institute of Structural Biology in Grenoble, Dr. Drouet's research investigates persisting viruses in humans (RNA and DNA viruses) and the balance with our host immune system. He focuses on these viruses' effects on humans (both their impact on pathology and their symbiotic relationships in humans). He has an excellent track record in the herpesvirus field, and his group is engaged in clinical research in the field of Epstein-Barr virus diseases. He is the editor of the online *Encyclopedia of Environment* and he coordinates the Universal Health Coverage education program for the BioHealth Computing Schools of the European Institute of Science.

Editor of Volume 11:
Emmanuel Drouet
Institut de Biologie Structurale
(Complement, antibodies and infectious diseases group)
Université de Grenoble-Alpes, Grenoble, France

Book Series Editor:
Shailendra K. Saxena
King George's Medical University

Scope of the Series

The series will give a most comprehensive overview of recent trends in various infectious diseases (as per the most recent Baltimore classification), as well as general concepts of infections, immunopathology, diagnosis, treatment, epidemiology and etiology to current clinical recommendations in management of infectious diseases, highlighting the ongoing issues, recent advances, with future directions in diagnostic approaches and therapeutic strategies. This book series will focus on various aspects and properties of infectious diseases whose deep understanding is very important for safeguarding human race from more loss of resources and economies due to pathogens.

Contents

Preface

Epstein-Barr virus (EBV) is one of the most common human viruses and the cause of pathologies such as infectious mononucleosis (IM) and certain cancers, namely immunodeficiency-related B cell lymphomas, Burkitt and Hodgkin, nasopharyngeal, and gastric carcinomas. Over the past two decades, the possibility of an association between EBV and other cancers and other chronic pathologies (i.e., multiple sclerosis (MS)) has also been put forward. One of the challenges facing researchers is the complicated life cycle of EBV, which goes through a phase of latent infection during which the virus induces the activation, proliferation, and differentiation of primary B cells into memory B cells. Additionally, EBV, like other human herpesviruses (HHV1-8), has co-evolved through a persistent viral infection in the host, and is then spread efficiently to others, generally without causing serious diseases. Symptoms of EBV infection vary widely based on the age and immune status of the patient. Most infections in younger children are benign and are often subclinical. EBV is also associated with autoimmune diseases, including rheumatoid arthritis, Sjogren's syndrome, systemic lupus erythematosus, and MS.

Classified as a herpesvirus (type IV), EBV encodes more than 80 genes. The core set of genes (minority) are involved in the latency phase. The other set comprises the genes of the lytic cycle. In addition to these gene-encoding proteins, there are gene-encoding microRNAs (regulatory RNAs), the functions of which are still poorly understood. For many years, researchers argued that only the products of the latency genes (i.e., LMP1 oncoprotein and EBNAs) were responsible for oncogenesis. It has recently been demonstrated that the proteins of the lytic cycle have also a role not only in cell transformation (the initial stage of the tumor process) but also in tumor progression. Certain viral proteins act as "transcription factors" capable of activating cellular genes involved in the regulation of cell survival or even in immunomodulation.

EBV-associated lymphomas are classically described as malignant proliferations of the lymphoid type but nonetheless group together a wide variety of histological and immunological types. In addition, this association with EBV, considered to be a group 1 carcinogen according to the International Agency for Research on Cancer (IARC) 2009, is highly variable for the type of lymphoma considered. For example, in Burkitt's lymphoma (BL), which was the first cancer associated with an infection and observed in Ugandan children thanks to the work of Denis Burkitt in 1958, we find it is the B lymphoma associated with EBV (discovered eight years later) that is present in more than 90% of cases. In contrast, in the same type of lymphoma but observed in a European subject, the association is only around 20%. This demonstrates that, in addition to the virus, there are environmental co-factors linked to each form of lymphoma, (regardless of whether it is type B or type T) or even a specific lymphoma, such as lymphoma by Hodgkin. The immune state is one of these cofactors linked to the host, and this explains the appearance of severe lymphoproliferative diseases in immunocompromised subjects (e.g., transplant recipients). In these patients, these lymphomas are called post-transplantation lymphoproliferative syndromes (PTLDs), bringing together several types of lymphomas (B lymphomas most often associated with EBV, but also T lymphomas, BLs, or Hodgkin's disease).

The chance of developing a lymphoma hangs over the heads of humans much like the sword of Damocles. Fortunately, we all have an internal immunosurveillance network capable of monitoring every cell infected with EBV. Over the long course of evolution spanning millions of years, a balance has been established between the virus and its host (primates and humans) so that we can live within this complex equilibrium. After initial exposure and infection with EBV (this frequently occurs in children and is asymptomatic), the virus persists in a form of latency with reactivations (periodic resumption of the activity of the virus, i.e., lytic cycle) that may go unnoticed. Under certain circumstances, the virus exerts uncontrolled oncogenic activity, which can sequentially (multi-step) lead to symptomatic tumors and cancer, such as lymphomas, stomach cancer (see Chapter 3), nasopharyngeal carcinoma (NPC), and leiomyosarcoma (LMS). For some of these tumors, EBV coinfection with malaria (African LB), ethnic and diet factors (NPC), and immune status (post-transplant lymphomas and lymphomas of the HIV-positive subject) are conducive to their development.

Recent studies have revealed a seamlessness between latent and lytic proteins and the types of infections to which they contribute. Some lytic proteins can be expressed in the context of latent infections as seen in some cancers and in pre-latent infection of B lymphocytes. This raises the possibility that these lytic antigens (specifically the ZEBRA protein encoded by the EBV BZLF1 gene) might be useful therapeutic or vaccine targets for the prevention of EBV-induced cancers. In addition, advances in high-throughput next-generation DNA sequencing have made it possible to analyze a growing number of EBV isolates (see Chapters 1 and 2). It appears that different isolates of EBV vary in their ability to infect lymphocytes and epithelial cells. It can thus be suggested that certain specific variants of EBV are more oncogenic than others, and we can therefore establish a clear link between EBV-induced oncogenesis and that of human papillomavirus (HPV). It is also plausible that a particular variant is more prone to reactivate towards the lytic cycle, causing an increased viral load or an increase in the circulating ZEBRA antigen. This protein could act as a tumor progression factor and increase the occurrence of EBV-induced cancers. Recent clinical studies demonstrate a causal role of EBV in MS and in myasthenia gravis (an autoimmune disease characterized by intrathymic B-cell activation; see Chapter 5). Once again, it is not yet clear why MS only develops in a small fraction of people infected with EBV. The role of gene variants is worth investigating in future studies.

Usually, the diagnosis of an EBV primary infection is established through serological testing (detection of antibodies specific to EBV). This is particularly so regarding EBV whose symptoms are found in adolescents suffering from IM. In certain cases (e.g., NPC), this serology can also be applied to consolidate a diagnosis or a follow-up therapy. In contrast, in the case of lymphomas, clinicians have chosen molecular biology techniques (viral load by quantification of circulating EBV DNA in the blood). This technique (exploring the virus in its latency form) is routinely used to monitor transplant patients and to intervene at an early stage with drugs (rituximab, an anti-B lymphocyte monoclonal antibody) to prevent the onset of PTLD. Unfortunately, this technique (Rituximab or its derivatives) has many side effects (among them, hypogammaglobulinemia) and therefore lacks specificity as a potential treatment. However, techniques based on the detection of the lytic cycle have been developed to improve the specificity of the diagnosis of PTLD and will soon be available to everyone.

Despite the discovery of EBV more than 50 years ago, immune control of the virus is not very well understood and there is still no vaccine available. This knowledge gap is due, in part, to the lack of a preclinical small animal model that can realistically recapitulate EBV infection and immune control and therefore allows testing of EBV-specific vaccine candidates (see Chapter 6). With the advent of mice with reconstituted human immune system compartments during the past decade, this is now changing. The complex interplay between host and virus has also made it difficult to elaborate useful vaccine strategies to protect against EBV-associated diseases (including chronic diseases like MS) or to find efficient drugs that specifically target EBV malignancies. Recently, the incorporation of immunotherapeutic strategies as first-line therapy has provided a better long-term outcome for patients. On the other hand, new predictive biomarkers have been found for patient follow-ups.

EBV is present in many pathologies, thus there is a need to encourage further research in this domain, which could lead to the discovery and development of new specific therapies.

Emmanuel Drouet, PharmD, Ph.D.
Full Professor,
Institut de Biologie Structurale,
(Complement, Antibodies and Infectious Diseases Group),
Université de Grenoble-Alpes,
Grenoble, France

Chapter 1

Genome-Wide Profiling of Epstein-Barr Virus (EBV) Isolated from EBV-Related Malignancies

Ying Liu, Zheming Lu and Hongying Huang

Abstract

Epstein–Barr virus (EBV) is the cause of certain cancers, such as Burkitt lymphoma, Hodgkin lymphoma, NK/T cell lymphoma, nasopharyngeal carcinoma, and a subset of gastric carcinomas. The genome-wide characteristics of EBV are essential to understand the diversity of strains isolated from EBV-related malignancies, provide the first opportunity to test the general validity of the EBV genetic map and explore recombination, geographic variation, and the major features of variation in this virus. Moreover, understanding more about EBV sequence variations isolated from EBV-related malignancies might give important implications for the development of effective prophylactic and therapeutic vaccine approaches targeting the personalized or geographic-specific EBV antigens in these aggressive diseases. In this chapter, we will mainly focus on the EBV genome-wide profiling in three common EBV-related cancers in Asia, including nasopharyngeal carcinoma, EBV-associated gastric carcinoma, and NK/T-cell lymphoma.

Keywords: Epstein–Barr virus (EBV), next-generation sequencing (NGS), nasopharyngeal carcinoma (NPC), EBV-associated gastric carcinoma (EBVaGC), NK/T-cell lymphoma (NKTCL)

1. Introduction

Epstein–Barr virus (EBV), a ubiquitous human herpesvirus discovered in 1964 is classified as a group I carcinogen by the International Agency for Research on Cancer (IARC), since the latent infection by EBV has been estimated to be responsible for 200,000 cancer cases worldwide [1], including Burkitt lymphoma, Hodgkin lymphoma, NK/T cell lymphoma (NKTCL), nasopharyngeal carcinoma (NPC), and a subset of gastric carcinomas. It has been shown that viruses can contribute to the biology of multistep oncogenesis and are implicated in many of the hallmarks of cancer [2]. Notably, the discovery of links between viral infection and cancer types has provided actionable opportunities, such as the use of human papilloma virus (HPV) vaccines as a preventive measure, to reduce the global impact of cancer. However, until now, approved vaccines for EBV have not been available.

EBV has a double stranded DNA genome comprised of approximately 172 kilobases. The expression products cover at least 86 proteins and 46 functional small-untranslated RNAs [3–5]. EBV has two distinct life cycles: latency and lytic replication. During latency, viral genomes only express a limited number of latent proteins (EBV-determined nuclear antigen 1 (EBNA1), 2, 3A, 3B, and 3C and EBNA leader protein (EBNA-LP); latent membrane protein 1 (LMP1) and LMP2 (which encodes two isoforms, LMP2A and LMP2B)), noncoding EBV-encoded RNAs (EBER1 and EBER2), and viral miRNAs (BHRF1-miRNA and BART-miRNA). EBV latency is categorized as three latency types (latency I–III). EBV genomes in type-I latency are known to express EBNA1 and EBER. EBV genomes in type-II latency are known to express more genes such as *EBNA-LP*, *LMP1*, *LMP2A*, and *LMP2B*. EBV genomes in type-III latency are known to express most restricted latent genes including *EBNA2*, *EBNA3A*, *EBNA3B*, and *EBNA3C*. Lytic genes encode viral transcription factors (e.g., BZLF1), a viral DNA polymerase (BALF5) and associated factors, and viral glycoproteins (e.g., gp350/220 and gp110) and structural proteins (capsid and tegument proteins).

Southern blot of restriction fragment length polymorphisms was first used to detect EBV strain variation, and Sanger sequencing of certain specific viral genes (e.g., EBNA1 and LMP1) was later developed to detect sequence diversity. Now, on the basis of high-throughput sequencing, genome-wide analysis is becoming possible.

Prior to 2013, EBV whole genome sequences available from GenBank were limited to less than 10 strains (B95-8, EBV-WT, GD1, AG876, GD2, HKNPC1, Akata, and Mutu). The prototypic type 1 EBV strain B95-8 was the first complete genome sequenced from an individual with infectious mononucleosis using a conventional strategy (i.e., subcloning followed by Sanger sequencing) [6]. Subsequently, a more representative type 1 EBV reference genome, human herpesvirus 4 complete wild type genome, was constructed by using B95-8 as the backbone with an 11-kb deletion segment provided by the Raji sequences (named EBV-WT) [7]. AG876 was the unique complete type 2 EBV sequence from a Ghanaian case of Burkitt lymphoma [8]. Akata and Mutu were sequenced from Burkitt lymphoma cell lines from a Japanese patient and a Kenyan patient, respectively [9]. GD1 [10], GD2 [11] and HKNPC1 [12] were isolated from NPC patients.

Since 2014, a new technology named Hybrid Capture (**Figure 1**), has marked a new era of EBV genome sequencing. Using the method of target enrichment of EBV DNA by hybridization, followed by next-generation sequencing, de novo assembly, and joining of contigs can yield complete EBV genomes. The development of high-throughput sequencing technologies enabled sequencing of EBV genomes derived from a wide variety of clinical samples, such as tumor biopsy samples [13]. The number of available EBV sequences is increasing exponentially and up to now, more than 500 EBV genomes have been sequenced from a variety of human malignancies, including NPC, lymphoma, gastric cancer, and lung cancer, as well as from healthy carriers [14–25]. Progress has made it possible that the population-based case–control studies of EBV strain variation in EBV-related cancer patients as compared with the healthy population and a comprehensive survey of EBV integration in a variety of human malignancies can be effectively conducted [20, 25–27]. These developments have revealed that various EBV strains are differentially distributed throughout the world, and that the behavior of cancer-derived EBV strains is different from that of the prototype EBV strain of noncancerous origin.

Hence, the genome-wide characteristics of EBV are essential to assess the diversity of strains isolated from EBV-related malignancies. Meanwhile, understanding the pattern of EBV sequence variation is important for knowing whether there is a disease-related strain-specific or geographic regional variation of EBV strain, and might provide important implications for the development of effective prophylactic and therapeutic vaccine approaches targeting the personalized or geographic-specific EBV antigens in these aggressive diseases.

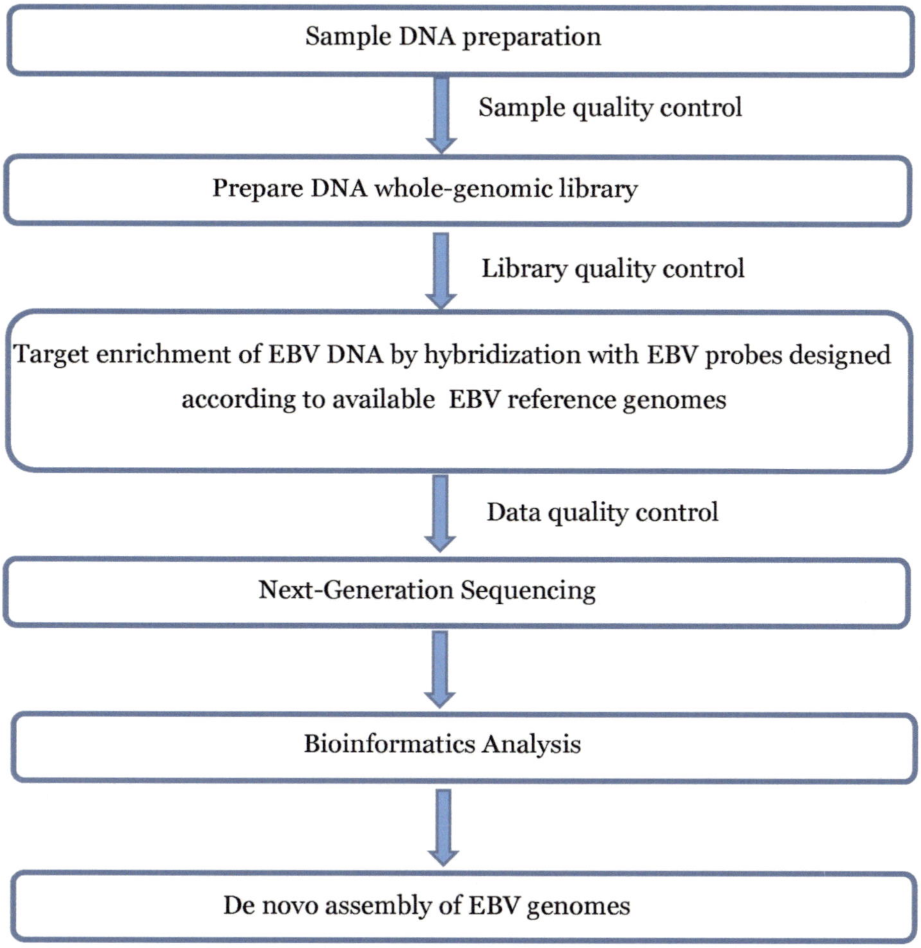

Figure 1.
Complete workflow for EBV DNA capture and sequencing.

EBV strain	Origin	Disease	Year	Refs
GD1	Guangdong, China	NPC	2005	[10]
GD2	Guangdong, China	NPC	2011	[11]
HKNPC1	Hong Kong, China	NPC	2012	[12]
HKNPC2-HKNPC9	Hong Kong, China	NPC	2014	[14]
EBVaGC1-EBVaGC9	Beijing, China	EBVaGC	2016	[17]
GDGC1-GDGC2	Guangdong, China	EBVaGC	2018	[21]
NKTCL-EBV1-NKTCL-EBV8	Beijing, China	NKTCL	2019	[23]
NKTCL-SC01-NKTCL-SC15	15 from Southern China, 12 from Singapore	NKTCL	2019	[24]
NKTCL-SG01-NKTCL-SG12				

Table 1.
EBV genomes reviewed in this chapter

In this chapter, EBV genomes reviewed are from three common EBV-related cancers in Asia, including NPC, EBV-associated gastric carcinoma (EBVaGC), and NKTCL. The EBV strains include GD1 [10], GD2 [11], HKNPC1 [12],

HKNPC2-HKNPC9 [14], EBVaGC1-EBVaGC9 [17], GDGC1-GDGC2 [21], NKTCL-EBV1-NKTCL-EBV8 [23], NKTCL-SC01-NKTCL-SC15 and NKTCL-SG01-NKTCL-SG12 [24] (**Table 1**).

2. Genomic diversity of EBV-related malignances

2.1 NPC

NPC, an EBV-associated epithelial carcinoma, has a unique geographical distribution [28]. A recent World Health Organization (WHO) report estimated that there were around 130,000 new NPC cases worldwide in 2018 [29]. Rare in most of the world, NPC is particularly prevalent in South China and Southeast Asia [30]. In Hong Kong and Guangdong in South China, NPC incidence is as high as 12.8–25.0/100,000 per year [28, 29]. The cause of NPC endemicity remains unknown.

Many studies have shown that EBV genome is present in almost all endemic NPC tumors with a unique pattern of virus latent gene expression, suggesting that EBV plays an important role in the tumorigenesis of NPC [31]. Whole genome sequencing is useful for us to understand genomic characterization and divergence. Here, we mainly focus on 11 mostly available full-length genomes of NPC.

2.1.1 GD1

GD1 (Guangdong strain 1), the first NPC-derived EBV strain with full-length sequences determined using PCR amplification and sub-cloning followed by conventional Sanger sequencing technology, was analyzed from established a lymphoblastoid cell line (LCL) from umbilical cord blood mononuclear cells transformed by saliva virus from a Cantonese NPC patient in 2005 [10]. The entire GD1 sequence is 171,656 bp in length and GD1 belongs to type 1 strain. Many sequence variations in GD1 compared to prototypical strain B95-8 were detected, including 43 deletion sites, 44 insertion sites, and 1413 point mutations. Furthermore, the frequency of some GD1 mutations in Cantonese NPC patients was evaluated, such as a 30-bp deletion in the C terminus of LMP1, and the V-Val subtypes of EBNA1. The results suggested that GD1 is highly representative of the EBV strains isolated from NPC patients in Guangdong, China, an area with the highest incidence of NPC in the world.

2.1.2 GD2

With the invention of next-generation sequencing (NGS) systems, it is possible to determine genome-wide sequences and the viral clonality of EBV strains by direct sequencing of EBV genomes in clinical tumors in a time- and cost-effective manner. GD2 with 164,701 bp long was directly sequenced using the Illumina (Solexa) platform, and successfully assembled from an NPC tumor of a patient in Guangdong province, a region in China by the same group who determined GD1 [11]. GD2 was closely related to GD1 by sequence and phylogenetic analyses. The sequence similarities between GD2 and GD1 were 98.76%. GD2 and GD1 shared 505 common single-nucleotide variations (SNVs), including most SNVs in the coding regions (348 [68.91%] SNVs) and seven insertion and deletions (indels). From a comparison with the EBV-WT reference genome, a total of 927 SNVs and 160 indels with genome-wide distribution were found in the GD2 genome. The results revealed that NGS allows the characterization of genome-wide variations of EBV in clinical tumors and provides evidence of monoclonal expansion of EBV *in vivo*.

2.1.3 HKNPC1

Because of the relatively small quantity of viral DNA present in the tumor sample, next-generation sequencing total cellular and viral DNA in a sample is costly and inefficient, and may limit the generation of the high read depth necessary to make high confident base calls of the viral genome. Using target enrichment technology could increase the relative amount of viral DNA. Kwok et al. reported an approach of PCR enrichment (Amplicon Sequencing) followed by sequencing the amplified products on the Illumina Genome Analyzer IIx platform to determine the genome sequence of an EBV isolate from NPC tumor of a Chinese patient in Hong Kong, designated as HKNPC1 [12]. HKNPC1 is approximately 171,549 bp, and contains 1589 SNVs and 132 indels in comparison to the reference EBV-WT sequence. Non-synonymous SNVs were mainly found in the latent, tegument and glycoprotein genes. The same point mutations were found in glycoprotein (BLLF1 and BALF4) genes of GD1, GD2 and HKNPC1 strains and might affect cell type specific binding. The results showed that whole genome sequencing of EBV in NPC may facilitate discovery of previously unknown variations of pathogenic significance.

2.1.4 HKNPC2-9

The group of Kwok and colleagues established a complete sequencing workflow comprising target enrichment of EBV DNA by hybridization, followed by next-generation sequencing, *de novo* assembly, and joining of contigs by Sanger sequencing to yield whole EBV genomes. The sequences of eight NPC biopsy specimen-derived EBV (NPC-EBV) genomes, designated HKNPC2 to HKNPC9, were then determined in the same geographic location in order to reveal their sequence diversity [14]. The eight NPC-EBV genome sizes estimated based on the reference EBV-WT sequence ranged from 170,062 bp (HKNPC2) to 171,556 bp (HKNPC3 and -6). A total of 1736 variations were found, including 1601 substitutions, 64 insertions, and 71 deletions, compared to the reference EBV-WT genome. Furthermore, genes encoding latent, early lytic, and tegument proteins and glycoproteins were found to contain nonsynonymous mutations of potential biological significance. Thus, much greater sequence diversity among EBV isolates derived from NPC biopsy specimens is demonstrated on a whole-genome level through a complete sequencing workflow.

Obtaining whole-genome sequence information for more clinical EBV isolates, with good representation of the EBV repertoire in tumors, could help to address that hypothesis and uncover the pathogenic subtypes of EBV in NPC tumorigenesis. A case–control (62 NPC patients and 142 population carriers) study of NPC in Hong Kong has identified high-risk EBV subtypes with polymorphisms in the EBV-encoded small RNA (EBER) locus [26]. A recent study published in Nature Genetics entitled 'Genome sequencing analysis identifies high-risk Epstein–Barr virus subtypes for nasopharyngeal carcinoma' by Xu et al. used large-scale EBV whole-genome sequencing to examine EBV subtypes in an attempt to explain the unique NPC endemicity in South China [25]. Through EBV genomes from 156 NPC cases and 47 controls and two-stage association study, they identified two non-synonymous EBV variants within the BALF2 gene (BamHIA leftward reading frame 2 encoding a single strand DNA binding protein associated with EBV replication) strongly associated with the risk of NPC (odds ratio [OR] = 8.69 for SNP162476_C and OR = 6.14 for SNP163364_T). The cumulative effects of these variants contribute to 83% of the overall risk of NPC in southern China. These studies confirmed the critical role of EBV infection in the pathogenesis of NPC and provided an explanation for the striking epidemiological distribution of this tumor in South China.

2.2 EBVaGC

EBVaGC has been recognized as a distinct subset of gastric carcinoma, accounting for about 10% of total gastric carcinomas [32–35]. The monoclonal presence of the virus was uniformly distributed in malignant cells of EBV-positive tumors but not observed in the surrounding normal epithelial cells, providing strong evidence to support the role of EBV as an etiologic agent [32, 33]. However, the exact role of EBV in the development and progression of this specific type of gastric carcinoma is not yet clear.

Progress has been made in understanding the full spectrum of diversity existent within the EBV genome from EBVaGC clinical tumor samples, since the NGS technology has been developed. Here, 11 EBV strains from primary EBVaGC biopsy samples were included.

2.2.1 EBVaGC1-EBVaGC9

Our group reported the first genome-wide view of sequence variation of EBV isolated from primary EBVaGC biopsy specimens in 2016 [17]. We used the method of target enrichment of EBV DNA by hybridization, followed by next-generation sequencing. EBV probes were designed according to full-length genome of six available EBV strains, including EBV-WT, B95-8, AG876, GD1, GD2, and HKNPC1. According to the value of coverage of the target region, all DNA sequence generated from GC-EBV strains most resembled GD1. Thus, GD1 was used as the reference EBV genome in our study. *De novo* assembly was performed for nine sequenced GC-EBV strains. Finally, nine EBVaGC genomes were successfully sequenced, designated EBVaGC1 to EBVaGC9. The genome sizes, estimated based on the reference GD1 sequence, ranged from 171,612 bp (EBVaGC6) to 171,957 bp (EBVaGC1).

Whole-genome sequencing of EBV enabled the comparison and thus the determination of EBV variations at the genome level. In our study, 961 variations were observed in the EBVaGC1 to 9 genomes in comparison to the reference GD1, including 919 substitutions, 23 insertions, and 19 deletions. Both latent genes and genes encoding tegument proteins in nine GC-EBV genomes were found to harbor the majority of nonsynonymous mutations, accounting for 58.4% (EBVaGC8) to 84.3% (EBVaGC3) of all nonsynonymous mutations detected for each genome.

EBNA1 is essential for maintenance of the EBV episome in latently infected cells and is the only EBV antigen that is consistently expressed in all EBV associated malignancies [36]. Based on the amino acid changes at position 487 in the COOH-terminal region in EBNA1 relative to B95-8 (P-ala), V-val was the most common subtype, accounting for 77.7% of nine GC-EBV strains, followed by P-thrV, accounting for 22.3%. Multiple results showed that V-val is the dominant subtype in Asian regions studies, not only in EBVaGC but also in NPC and healthy donors, while V-val subtype was rarely found in Africa, Europe, and America irrespective of source (lymphoma, NPC, EBVaGC, or healthy donors) [37–39], indicating that polymorphism of EBNA1 subtypes has geographic differences but is not tumor-specific. Apart from changes in the C-terminus, EBNA1 has variations in the N-terminus. Interestingly, we identified two interstrain recombinants at the EBNA1 locus, which provided a further mechanism for the generation of diversity. EBNA1 N-terminus changes have revealed additional variants that were not simply classified based on the signature amino acid residue 487 in the C-terminus as widely used previously. The N-terminus changes reinforce the need to evaluate the EBV genome more comprehensively in order to characterize the full extent of EBV genetic diversity. A comprehensive investigation into the functional and immunological impact of the naturally occurred *EBNA1* sequence variations and interstrain recombinants is required to evaluate their possible significance, which may also be helpful for clarifying the association of EBNA1 subtypes and EBVaGC.

2.2.2 GDGC1 and GDGC2

In 2018, NGS was employed to determine the EBV genomes from two EBVaGC specimens, designated as GDGC1 and GDGC2, from Guangdong, China, an endemic area of NPC [21]. Due to the presence of the much more abundant cellular genomic DNA in the DNA preparations, the number of reads belonging to EBV was low, accounting for only 0.02–0.23% of the total reads. However, since the original data were sufficient, the average sequencing depth for genomes GDGC1 and GDGC2 was ~73x and ~24x, respectively, which was sufficient for further analysis. The genome sizes, estimated based on the reference EBV-WT genome sequence, were as follows: GDGC1 (169,611 bp) and GDGC2 (171,299 bp).

The authors reported that a total of 1815 SNPs (146 indels) and 1519 SNPs (106 indels) were found in GDGC1 and GDGC2, respectively, compared with the reference EBV-WT genome. Among these, 1229 SNPs (66 indels) and 1076 SNPs (54 indels) were located in the coding regions for GDGC1 and GDGC2, respectively, while the remaining variations were found in the non-coding regions. Consistent with previous reports [17], there is clear evidence for a higher frequency of SNPs in latent genes, followed by the genes encoding tegument and membrane glycoproteins. In contrast to the frequent mutations that occurred in latent genes, the sequences of promoters and ncRNAs were investigated to be strictly conserved. A few point mutations were found in the sequences of Cp, Qp, Fp and LMP2Ap, and only scattered mutations could be identified in certain ncRNA sequences. Promoters and EBV-generated ncRNAs play important roles in regulating viral processes and in mediating host-virus interactions. Thus, a detailed EBV genome-wide analysis of EBVaGC from Guangdong was performed, which would be helpful for further understanding of the relationship between EBV genomic variation and EBVaGC carcinogenesis.

The features of the disease and geographically associated EBV genetic variation as well as the roles that the variation plays in carcinogenesis and evolution remain unclear. A recent study sequenced 95 geographically distinct EBV isolates from EBVaGC biopsies (n = 41) and saliva of healthy donors (n = 54) to detect variants and genes associated with gastric carcinoma from a genome-wide spectrum [20]. BRLF1, BBRF3, and BBLF2/BBLF3 genes had significant associations with gastric carcinoma. LMP1 and BNLF2a genes were strongly geographically associated genes in EBV. The results provided insights into the genetic basis of oncogenic EBV for gastric carcinoma, and the genetic variants associated with gastric carcinoma could serve as biomarkers for oncogenic EBV.

2.3 NKTCL

Extranodal NKTCL, a rare type of non-Hodgkin lymphoma, is characterized by the presence of EBV in virtually all cases, irrespective of their ethnicity or geographical origin. NKTCL is an aggressive malignancy, predominantly occurs in the nasal, paranasal, and oropharyngeal sites, and is much more prevalent in East Asia and Latin America than in Western countries [40].

Although the association of this B lymphotropic virus with malignancies of T and NK cell origin was quite unexpected, both the presence of virus sequences in tumor cells and the virus's oncogenic potency have led to the hypothesis that whether particular EBV strains are preferentially selected in NKTCL. Pathogenesis and genotype analyses of NKTCL have mainly focused on genetic variations in a small fraction of EBV genes before, which is limited to define the spectrum of diversity within the whole genome of EBV. The genome-wide characteristics of EBV are essential to understand the diversity of strains isolated from NKTCL. In 2019, for the first time, 35 NKTCL-derived EBV genomic landscapes at genome-wide level were simultaneously systematically characterized by two groups.

2.3.1 NKTCL-EBV1-NKTCL-EBV8

Our group directly sequenced EBV-captured DNA from eight primary NKTCL biopsy samples from China using Illumina HiSeq 2500 sequencer platform and presented the eight EBV sequences, designated NKTCL-EBV1-NKTCL-EBV8 [23]. Aiming at knowing the detail of subtype, the obtained DNA sequences were compared with six mostly referenced sequences, including AG876, B95-8, EBV WT, GD1, GD2, and HKNPC1. The GD1 coverage percentages are higher than the rest. The genome sizes, estimated based on the reference GD1 sequence, ranged from 171,590 bp (NKTCL-EBV8) to 172,059 bp (NKTCL-EBV1).

Whole-genome sequence alignments revealed extensive nucleotide variation in the eight NKTCL-EBV genomes. In comparison with the most similar GD1 strain, the NKTCL-EBV1 to NKTCL-EBV8 harbored 2072 variations in total, including 1938 substitutions, 58 insertions, and 76 deletions. Among them, 1218 substitutions, 15 insertions, and 26 deletions were located in the coding regions. The number of the nonsynonymous mutations is highest in the gene regions encoding latent proteins in each of the NKTCL-EBV genomes, followed by genes encoding the tegument protein and membrane glycoproteins.

EBNA1 and LMP1 are the most frequently studied regions to date. Based on the amino acid changes in certain residues of LMP1 and EBNA1, eight NKTCL-EBVs were sorted to China 1 and V-val subtype, respectively. Of interest, EBNA1 of NKTCL-EBV3 sequence showed clustered away from the other seven NKTCL-EBV strains. Analysis of amino acid sequences of EBNA1 supported that EBNA1 of NKTCL-EBV3 may arise from recombination of GD1 and B95-8. Other two commonly classification systems for LMP-1 gene polymorphisms include a 30-bp deletion in the C terminus and the loss of restriction site Xho I in the N terminus of the gene. LMP1 is a key latent protein with abilities to promote cell proliferation and inhibit cell apoptosis in NKTCL. In our study, the LMP1 strain in NKTCL-EBV1-NKTCL-EBV7, but not NKTCL-EBV8, harbored the 30-bp deletion. The variant of 30-bp deletion of LMP1 has been demonstrated that it is associated with poor prognosis of patients with NKTCL, which might serve as a potential marker to monitor treatment [41]. In addition, eight NKTCL-EBV strains had Xho I restriction site loss at exon 1 of the LMP1 gene.

2.3.2 NKTCL-SC01-NKTCL-SC15 and NKTCL-SG01-NKTCL-SG12

The other group assembled 27 NKTCL-derived EBV genome sequences retrieved from whole-genome sequencing data using the Hiseq sequencer (Illumina), including 15 EBV-positive NKTCL tumor samples from Southern China and 12 samples from Singapore [24]. The average percentage of EBV sequences in WGS data is 0.45% (0.03–1.06%), and the coverage depth is 222.2X in average (26.7X–612.8X). As ~34 kb of 172 kb of EBV genome are repeat regions, which could not be properly assembled with short-reads sequencing technology, the groups assigned "N" for these regions and subsequently joined the scaffolds, resulting in EBV genomes with ~172 kb in length.

The authors reported that among the 27 NKTCL samples, in average 1152 EBV SNVs for each sample were determined by aligning the viral reads against the reference EBV-WT genome. The most frequent tumor-specific non-synonymous mutations in NKTCL-derived EBV were located at BPLF1 gene (position 49,790–59,239 bp). An average of 44.8 small indels (<50 bp) of EBV were found in each NKTCL sample, and the 30-bp deletion of LMP1 was commonly found in the samples (21/27), with a frequency consistent with the previous study revealed by using Sanger sequencing [42]. Large deletions of EBV (>1 kb) were found in 10 of 27 NKTCL samples, without any sequencing coverage in the deleted regions. The findings provided insights into the understanding of EBV's role the etiology of NKTCL.

A genome-wide association study of 189 patients with extranodal NKTCL, nasal type and 957 controls from Guangdong province, Southern China was performed to identify common genetic variants affecting individual risk of NKTCL [43]. All cases were genotyped with Illumina Human OmniExpress ZhongHua-8 BeadChip, and population controls were scanned by Illumina OmniHumanExpress-24 V1.0 (both Illumina, San Diego, CA, USA). The findings were validated in four independent case–control series. The SNP with the strongest association was rs9277378 (OR 2.65 [95% CI 2.08–3.37]) located with HLA-DPB1, indicating the importance of HLA-DP antigen presentation in the pathogenesis of NKTCL. The pathogenic subtypes of EBV in NKTCL tumorigenesis should be further explored.

3. Phylogenetic analysis of the EBV genomes

Phylogenetic analysis of EBV genomes could demonstrate detailed overall genomic differences in EBV genome within or beyond subtypes of EBV-associated diseases, thus, EBV genomic similarity is likely to better infer the phylogenetic relatedness among EBV genomes.

Traditionally, EBV has two distinct subtypes, type 1 and type 2. Type 1 EBV (e.g., B95-8, GD1 and Akata) is the main EBV strain prevalent worldwide, while type 2 EBV (e.g., AG876) is abundant only in parts of Africa and New Guinea. Type 1 and type 2 EBV encode different *EBNA2* genes, with only 54% amino acid sequence identity. A recent whole genome sequencing study confirmed that *EBNA2* and *EBNA3* are the only genes that can distinguish type 1 and type 2 EBV strains [16]. Technologies for genome sequencing were currently developed with tools for genome analysis. High-throughput sequencing technology such as illumine dye sequencing was introduced to successfully sequence viral genomes. As exemplary tools for genome analysis, Molecular Evolutionary Genetics Analysis (MEGA) is used for both conducting statistical analysis of molecular evolution and constructing phylogenetic trees [44].

The NPC genomes from Asian EBV strains, including GD1, GD2, and HKNPC1-HKNPC9, are type 1 viruses and were clustered in a branch distant to the non-Asian strains AG876, B95-8 [14]. Analysis of LMP1 and -2 showed a phylogenetic relationship corresponding to the geographical origin of the viral genomes instead of the type 1 and 2 dichotomy, indicating that LMP1 and -2 genes can serve as geographical markers. GD1 seemed to harbor many mutations that were not present in the other Chinese strains. HKNPC6 and -7 genomes, which were isolated from tumor biopsy specimens of advanced metastatic NPC cases, were distinct from the other NPC-EBV genomes. Future work should investigate the relationship between the distinct lineage of EBV and the clinical stages of NPC.

GC-EBV strains, EBVaGC1-EBVaGC9 and GDGC1-GDGC2 involved here, were closely related to all Asian-derived EBV strains, distant to the non-Asian strains, and also showed that the EBV sequences generally clustered in a manner consistent with geographical location [17, 21]. Neighbor-joining trees derived from the sequences of *EBNA2* gene showed that all the GC-EBV genomes are type 1 viruses, clustered in a branch with other type 1 EBV strains, distant to the only type 2 EBV strain, AG876. Phylogenetic trees based on the LMP1 gene and whole EBV genomes indicated that the nine EBVaGC strains were closely related to all Asian-derived EBV strains and distant to the non-Asian strains, suggesting that the LMP1 gene can serve as a geographical marker [17]. This is in line with the previous results from the NPC-EBV genomes [14]. In addition, phylogenetic analyses on GDGC1 and GDGC2 derived from specific EBV-encoded gene suggested the presence of at least two parental lineages of EBV, as GDGC1 and GD2 clustered closely, while GDGC2 and GD1 clustered closely [21].

In our recent study, the phylogenetic trees were conducted based on alignment of eight full-length NKTCL-EBVs and previously published 28 strains [23]. Of note, eight NKTCL-EBVs genomes clearly sort into type 1, based on differences in whole genome and especially EBNA2. Eight NKTCL-EBVs were related to other Asian EBV strains, including EBVaGC1–9, HKNPC1–9, GD1, and GD2 obtained from China, and Akata from Japan, whereas none of the specimens was clustered

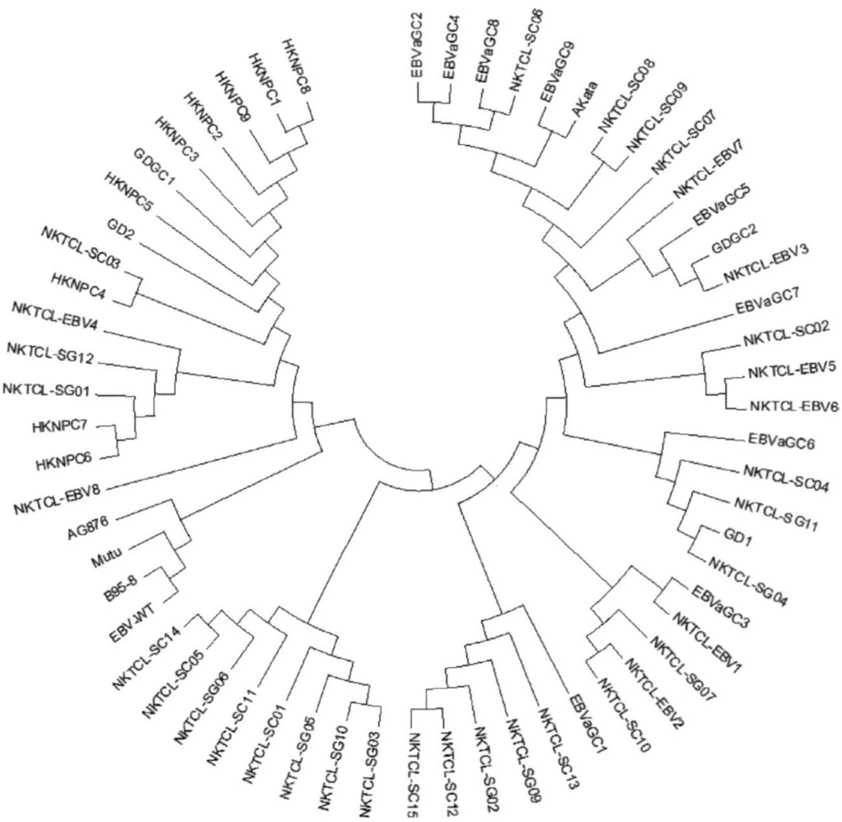

Figure 2.
Phylogenetic trees of EBV genomes. Phylogenetic analyses were conducted using the neighbor-joining (NJ) algorithm implemented in MEGA software (version 6). Bootstrap analysis of 1000 replicates was performed to determine the confidence.

in a branch of non-Asian strains AG876, B95-8, and Mutu. Other group compared the sequences between 27 NKTCL-derived EBV and 164 EBV genome sequences from public database to determine the sequence diversity of EBV [24]. Phylogenetic analysis revealed clear clustering of EBV isolates firstly according to their respective geographic origin; moreover, EBV isolates derived from NKTCL samples tend to cluster closely, apart from clusters by other diseases, supporting the hypothesis of the existence of disease-specific EBV. However, whether the unique EBV has been driving the development of NKTCL or simply adapted to the niche of NKTCL as bystander await further investigations.

In this chapter, phylogenetic analysis was conducted on full-length EBV genomes, including 11 NPC-EBV strains (GD1, GD2, HKNPC1-HKNPC9), 11 GC-EBV strains (EBVaGC1-EBVaGC9, GDGC1-GDGC2), 35 NKTCL-EBV strains (NKTCL-EBV1-NKTCL-EBV8, NKTCL-SC01-NKTCL-SC15, NKTCL-SG01-NKTCL-SG12), B95-8, EBV-WT, Mutu, Akata, and AG876 (**Figure 2**). The result of phylogenetic tree supports the conclusion that EBV infections are more likely affected by different geographic regions rather than particular EBV-associated malignancies.

4. Amino acid changes in CD4$^+$ and CD8$^+$ T-cell epitopes

Sequence variations of EBV genes also result in amino acid epitope exchanges, which should have a significant impact on EBV-specific T-cell immunity.

Among the shared non-synonymous SNVs of the Chinese derived GD1, GD2 and HKNPC1 isolates, 34 are associated with known EBV-specific epitopes; 19 and 15 are found in CD8$^+$ and CD4$^+$ epitopes, respectively [12]. HKNPC2-9 genomes harbored nonsynonymous mutations in epitopes specific for both CD4$^+$ and CD8$^+$ T cells [14]. Amino acid changes were found in seven CD8$^+$ epitopes of LMP2, five epitopes of EBNA3A, and three or fewer in other proteins. Thirteen CD4$^+$ epitopes of EBNA1, six in LMP1, six in LMP2, five in EBNA2, and three or fewer in other proteins contained amino acid changes. Some of the nonsynonymous mutations were affecting multiple epitopes.

EBVaGC shows EBV type I latency neoplasm, in which EBNA1 is expressed in 100% and LMP2A in about half of EBVaGC cases, respectively [45]. Recent studies show that EBNA1, as well as LMP2A, can be presented to both CD4$^+$ and CD8$^+$ T cells, highlighting its potential importance in the development of therapeutic strategies against EBV-associated malignancies [46, 47]. There is some clear evidence for sequence variation affecting immune recognition of EBNA1 and potential epitope selection for vaccine development [46]. So far, most research on the EBNA1 protein has been focused exclusively on the B95-8 strain alone [46, 47]. Sequence analysis of the gene encoding EBNA1 in EBV isolates from nine EBVaGC specimens has revealed considerable *EBNA1* sequence divergence from the B95-8 strain [17]. Importantly, T cell recognition of EBNA1 epitope might be greatly influenced by this sequence polymorphism as adoptive transfer of EBNA1-targeted T cells has a potential use in immunotherapy of EBV associated carcinomas.

NKTCL is associated with type II EBV latency, in which only restricted EBV antigens, namely EBNA1, and LMP1 and 2, are expressed [48]. These EBV encoded proteins might be the targets of immune recognition during its persistent infection, and their nonsynonymous variations in CD4$^+$ and CD8$^+$ T-cell epitopes may affect the efficacy for a cytotoxic T lymphocyte (CTL)-based therapy. Many epitopes were defined and were mapped in EBV antigens and correlated with major histocompatibility complex type in previous studies. In our study, we mainly investigated the amino acid changes in CD4$^+$ and CD8$^+$ T-cell epitopes of

EBNA1, LMP1, and LMP2A. Compared with B95-8, amino acids changes were found in 3 CD8$^+$ epitopes of EBNA1, 8 epitopes of LMP1, and 12 epitopes of LMP2A. Eleven CD4$^+$ epitopes of EBNA1, 13 in LMP1, and 9 in LMP2A contained amino acids. Some of the nonsynonymous mutations were affecting multiple epitopes [23]. In another study, alterations of the known T-cell epitopes were examined in EBV sequences derived from NKTCL [24]. Alterations of T-cell epitopes were detected in EBV derived from NKTCL samples. Notably, 21 of these epitopes with significant enrichment in NKTCL samples were restricted to six EBV genes, including EBNA3A (G373D, F325L, I333K, L406P, S412R, H464R, M466R, T585I, and A588P), EBNA3B (A399S, V400L, V417L, K424T, Y662D, and K663E), EBNA3C (P916S), BARF1 (V29A), BCRF1 (V6M), and BNRF1 (G456R, S497G, and A1289T).

Therefore, these data have implications for the development of effective prophylactic and therapeutic vaccine approaches targeting the personalized EBV antigens in these aggressive diseases. Adoptive transfer of cytotoxic T cells (CTLs) specific for EBV antigens has proved safe and effective as prophylaxis and treatment for EBV-associated lymphoproliferative disease. Some patients with advanced stage or relapsed EBV-associated malignancies achieved complete remission after treatment with autologous LMP1/2- and EBNA1-specific CTLs or activated by peptides derived from LMP1/2 [49, 50]. Nonetheless, some cases still did not respond to LMP-CTL therapy, and this failure was usually attributed to immune escapes by antigen loss. It is worth noting that all these previous studies used prototype EBV sequence, B95-8, to design full-length LMP epitopes. Therefore, recent work gives an alternative explanation for the lack of tumor response. Whether changes in such epitopes confer immune evasion of the tumor cells may constitute another hypothesis for future testing.

5. Genomic integration of viral sequences

Viral integration into the host genome has been shown to be a causal mechanism that can lead to the development of cancer [51]. Not surprisingly, known tumor-associated viruses, such as EBV, HBV, HPV16 and HPV18, were among the most frequently detected targets [52]. Notably, the approach of WGS is sensitive to detect viruses. This is particularly true for the common integration verified for HBV, HPV16 and HPV18 in a variety of studies [53–55]. The known causal role of HPV16 and HPV18 in several tumor entities, which triggered one of the largest measures in cancer prevention, has been the motivation for extensive elucidation of the pathogenetic processes involved. Integration events with high confidence were demonstrated for HBV (liver cancer), HPV16 and HPV18 (in both cervical and head-and neck carcinoma), however, low-confidence integration events were detected for EBV (gastric cancer and malignant lymphoma) [56].

Comprehensive analyses of WGS datasets may reveal some novel findings on EBV integration. Recently, a comprehensive survey of EBV integration in a variety of human malignancies, including NPC, EBVaGC, and NKTCL was conducted, using EBV genome capture combined with ultra-deep sequencing, which could efficiently detect integrated EBV sequences from background "noise" introduced by nuclear EBV episomes [27]. The EBV integration rates were 25.6% (10/39), 16.0% (4/25), 9.6% (17/177) in the EBVaGC, NKTCL, and NPC tumors, respectively, which were lower than HPV integration in cervical cancer (76.3%) and head and neck squamous cell carcinoma (60.7%), and HBV in hepatocellular carcinoma (92.6%) [54, 57–59]. They found that EBV integrations into the introns could decrease the expression of the inflammation-related genes, TNFAIP3, PARK2, and CDK15, in NPC tumors [27]. The EBV integration breakpoints were frequently at oriP or terminal repeats, and were surrounded by microhomology sequences, consistent with a mechanism for integration involving viral genome replication and microhomology-mediated recombination, which has an important role in the integration of other tumorigenic viruses, HBV and HPV [54, 59]. Meanwhile, researchers also observed integrations of short EBV fragments into human chromosomes, coincident with episomal EBV genomes in NKTCL, and showed that 31 EBV-host integration sites were detected from eight NKTCL samples, and enriched in the repeat regions of human genome, such as SINE, LINE, and satellite [24].

However, there are still few studies on EBV integration based on WGS technology. In addition, authors only selected some potential breakpoints to perform PCR and Sanger sequencing for validating. For example, Xu et al. randomly select 12 integrations from 197 breakpoints identified from NPC and other EBV-associated malignancies, and only 10 breakpoints were successfully validated [27]. As integration of EBV sequence into the host genome and the consequent disruption of the important host genes might represent a novel tumorigenesis mechanism in EBV associated malignancies, all the potential EBV integration breakpoints should be validated and biological function of host genes involved should be further conducted.

6. Summary

In conclusion, full-length EBV genomes isolated from primary NPC, EBVaGC, and NKTCL biopsy specimens have been successfully sequenced and the sequence diversity on a whole-genome level has been analyzed, although their pathogenesis remains to be clarified. Phylogenetic analysis has shown that all aforementioned NPC, GC, and NKTCL-EBV strains are type 1 EBV and close to other Asian subtypes, leading to the conclusion that EBV infections are more likely affected by different geographic regions rather than particular EBV-associated malignancies. In addition, sequence variations of EBV genes also result in amino acid epitope exchanges, which should have a significant impact on EBV-specific T-cell immunity. Recent data have provided optimization proposal for selecting EBV genome for treatment from individual patients or at least predominant strains prevalent in geographical regions instead of commonly used B95-8 genome. We acknowledge that further characterizations of the molecular events would provide more information on the exact mechanisms underlying their pathogenic potentials and clinical significance.

Acknowledgements

This work was supported by grants from National Natural Science Foundation of China (81903155 to Y.L.), and Beijing Municipal Natural Science Foundation (7202023 to Y.L.), and Beijing Hospitals Authority Youth Program (QML20181106 to Y.L.)

Conflict of interest

The authors declare no conflict of interest.

Author details

Ying Liu[1], Zheming Lu[2*] and Hongying Huang[3*]

1 Laboratory of Genetics, Key Laboratory of Carcinogenesis and Translational Research (Ministry of Education), Peking University Cancer Hospital and Institute, Beijing, People's Republic of China

2 Laboratory of Biochemistry and Molecular Biology, Key Laboratory of Carcinogenesis and Translational Research (Ministry of Education), Peking University Cancer Hospital and Institute, Beijing, People's Republic of China

3 Department of Pathology, New York University Langone Medical Center, New York, USA

*Address all correspondence to: luzheming@bjmu.edu.cn; hongying.huang@nyumc.org

References

[1] Parkin DM. The global health burden of infection-associated cancers in the year 2002. International Journal of Cancer. 2006;**118**(12):3030-3044

[2] Mesri EA, Feitelson MA, Munger K. Human viral oncogenesis: A cancer hallmarks analysis. Cell Host & Microbe. 2014;**15**(3):266-282

[3] Tarbouriech N, Buisson M, Geoui T, Daenke S, Cusack S, Burmeister WP. Structural genomics of the Epstein-Barr virus. Acta Crystallographica. Section D, Biological Crystallography. 2006;**62**(Pt 10):1276-1285

[4] Swaminathan S. Noncoding RNAs produced by oncogenic human herpesviruses. Journal of Cellular Physiology. 2008;**216**(2):321-326

[5] Chen SJ, Chen GH, Chen YH, Liu CY, Chang KP, Chang YS, et al. Characterization of Epstein-Barr virus miRNAome in nasopharyngeal carcinoma by deep sequencing. PLoS One. 2010;**5**(9):e12745

[6] Farrell PJ. Epstein-Barr virus. The B95-8 strain map. Methods in Molecular Biology. 2001;**174**:3-12

[7] de Jesus O, Smith PR, Spender LC, Elgueta Karstegl C, Niller HH, Huang D, et al. Updated Epstein-Barr virus (EBV) DNA sequence and analysis of a promoter for the BART (CST, BARF0) RNAs of EBV. The Journal of General Virology. 2003;**84**(Pt 6):1443-1450

[8] Dolan A, Addison C, Gatherer D, Davison AJ, McGeoch DJ. The genome of Epstein-Barr virus type 2 strain AG876. Virology. 2006;**350**(1):164-170

[9] Lin Z, Wang X, Strong MJ, Concha M, Baddoo M, Xu G, et al. Whole-genome sequencing of the Akata and Mutu Epstein-Barr virus strains. Journal of Virology. 2013;**87**(2):1172-1182

[10] Zeng MS, Li DJ, Liu QL, Song LB, Li MZ, Zhang RH, et al. Genomic sequence analysis of Epstein-Barr virus strain GD1 from a nasopharyngeal carcinoma patient. Journal of Virology. 2005;**79**(24):15323-15330

[11] Liu P, Fang X, Feng Z, Guo YM, Peng RJ, Liu T, et al. Direct sequencing and characterization of a clinical isolate of Epstein-Barr virus from nasopharyngeal carcinoma tissue by using next-generation sequencing technology. Journal of Virology. 2011;**85**(21):11291-11299

[12] Kwok H, Tong AH, Lin CH, Lok S, Farrell PJ, Kwong DL, et al. Genomic sequencing and comparative analysis of Epstein-Barr virus genome isolated from primary nasopharyngeal carcinoma biopsy. PLoS One. 2012;**7**(5):e36939

[13] Kwok H, Chiang AK. From conventional to next generation sequencing of Epstein-Barr virus genomes. Viruses. 2016;**8**(3):60

[14] Kwok H, Wu CW, Palser AL, Kellam P, Sham PC, Kwong DL, et al. Genomic diversity of Epstein-Barr virus genomes isolated from primary nasopharyngeal carcinoma biopsy samples. Journal of Virology. 2014;**88**(18):10662-10672

[15] Santpere G, Darre F, Blanco S, Alcami A, Villoslada P, Mar Alba M, et al. Genome-wide analysis of wild-type Epstein-Barr virus genomes derived from healthy individuals of the 1,000 genomes project. Genome Biology and Evolution. 2014;**6**(4):846-860

[16] Palser AL, Grayson NE, White RE, Corton C, Correia S, Ba Abdullah MM, et al. Genome diversity of Epstein-Barr virus from multiple tumor types and

normal infection. Journal of Virology. 2015;**89**(10):5222-5237

[17] Liu Y, Yang W, Pan Y, Ji J, Lu Z, Ke Y. Genome-wide analysis of Epstein-Barr virus (EBV) isolated from EBV-associated gastric carcinoma (EBVaGC). Oncotarget. 2016;7(4):4903-4914

[18] Wang S, Xiong H, Yan S, Wu N, Lu Z. Identification and characterization of Epstein-Barr virus genomes in lung carcinoma biopsy samples by next-generation sequencing technology. Scientific Reports. 2016;**6**:26156

[19] Tu C, Zeng Z, Qi P, Li X, Yu Z, Guo C, et al. Genome-wide analysis of 18 Epstein-Barr viruses isolated from primary nasopharyngeal carcinoma biopsy specimens. Journal of Virology. 2017;**91**(17):e00301-17

[20] Yao Y, Xu M, Liang L, Zhang H, Xu R, Feng Q, et al. Genome-wide analysis of Epstein-Barr virus identifies variants and genes associated with gastric carcinoma and population structure. Tumour Biology. 2017;**39**(10):1010428317714195

[21] Chen JN, Zhou L, Qiu XM, Yang RH, Liang J, Pan YH, et al. Determination and genome-wide analysis of Epstein-Barr virus (EBV) sequences in EBV-associated gastric carcinoma from Guangdong, an endemic area of nasopharyngeal carcinoma. Journal of Medical Microbiology. 2018;**67**(11):1614-1627

[22] Hong S, Liu D, Luo S, Fang W, Zhan J, Fu S, et al. The genomic landscape of Epstein-Barr virus-associated pulmonary lymphoepithelioma-like carcinoma. Nature Communications. 2019;**10**(1):3108

[23] Lin N, Ku W, Song Y, Zhu J, Lu Z. Genome-wide analysis of Epstein-Barr virus isolated from extranodal NK/T-cell lymphoma, nasal type. The Oncologist. 2019;**24**(9):e905–e913

[24] Peng RJ, Han BW, Cai QQ, Zuo XY, Xia T, Chen JR, et al. Genomic and transcriptomic landscapes of Epstein-Barr virus in extranodal natural killer T-cell lymphoma. Leukemia. 2019;**33**(6):1451-1462

[25] Xu M, Yao Y, Chen H, Zhang S, Cao SM, Zhang Z, et al. Genome sequencing analysis identifies Epstein-Barr virus subtypes associated with high risk of nasopharyngeal carcinoma. Nature Genetics. 2019;**51**(7):1131-1136

[26] Hui KF, Chan TF, Yang W, Shen JJ, Lam KP, Kwok H, et al. High risk Epstein-Barr virus variants characterized by distinct polymorphisms in the EBER locus are strongly associated with nasopharyngeal carcinoma. International Journal of Cancer. 2019;**144**(12):3031-3042

[27] Xu M, Zhang WL, Zhu Q, Zhang S, Yao YY, Xiang T, et al. Genome-wide profiling of Epstein-Barr virus integration by targeted sequencing in Epstein-Barr virus associated malignancies. Theranostics. 2019;**9**(4):1115-1124

[28] Chen YP, Chan ATC, Le QT, Blanchard P, Sun Y, Ma J. Nasopharyngeal carcinoma. Lancet. 2019;**394**(10192):64-80

[29] de Martel C, Georges D, Bray F, Ferlay J, Clifford GM. Global burden of cancer attributable to infections in 2018: A worldwide incidence analysis. The Lancet Global Health. 2020;**8**(2):e180–e190

[30] Mahdavifar N, Ghoncheh M, Mohammadian-Hafshejani A, Khosravi B, Salehiniya H. Epidemiology and inequality in the incidence and mortality of nasopharynx cancer in Asia. Osong Public Health and Research Perspectives. 2016;7(6):360-372

[31] Young LS, Yap LF, Murray PG. Epstein-Barr virus: More than 50 years old and still providing surprises. Nature Reviews Cancer. 2016;**16**(12):789-802

[32] Shibata D, Weiss LM. Epstein-Barr virus-associated gastric adenocarcinoma. The American Journal of Pathology. 1992;**140**(4):769-774

[33] Tokunaga M, Land CE, Uemura Y, Tokudome T, Tanaka S, Sato E. Epstein-Barr virus in gastric carcinoma. The American Journal of Pathology. 1993;**143**(5):1250-1254

[34] Takada K. Epstein-Barr virus and gastric carcinoma. Molecular Pathology. 2000;**53**(5):255-261

[35] Murphy G, Pfeiffer R, Camargo MC, Rabkin CS. Meta-analysis shows that prevalence of Epstein-Barr virus-positive gastric cancer differs based on sex and anatomic location. Gastroenterology. 2009;**137**(3):824-833

[36] Fukayama M, Ushiku T. Epstein-Barr virus-associated gastric carcinoma. Pathology, Research and Practice. 2011;**207**(9):529-537

[37] Gutierrez MI, Raj A, Spangler G, Sharma A, Hussain A, Judde JG, et al. Sequence variations in EBNA-1 may dictate restriction of tissue distribution of Epstein-Barr virus in normal and tumour cells. The Journal of General Virology. 1997;**78**(Pt 7):1663-1670

[38] Chang KL, Chen YY, Chen WG, Hayashi K, Bacchi C, Bacchi M, et al. EBNA-1 gene sequences in Brazilian and American patients with Hodgkin's disease. Blood. 1999;**94**(1):244-250

[39] Chen YY, Chang KL, Chen WG, Shibata D, Hayashi K, Weiss LM. Epstein-Barr virus-associated nuclear antigen-1 carboxy-terminal gene sequences in Japanese and American patients with gastric carcinoma. Laboratory Investigation. 1998;**78**(7):877-882

[40] Xiong J, Zhao W. What we should know about natural killer/T-cell lymphomas. Hematological Oncology. 2019;**37**(Suppl 1):75-81

[41] Halabi MA, Jaccard A, Moulinas R, Bahri R, Al Mouhammad H, Mammari N, et al. Clonal deleted latent membrane protein 1 variants of Epstein-Barr virus are predominant in European extranodal NK/T lymphomas and disappear during successful treatment. International Journal of Cancer. 2016;**139**(4):793-802

[42] Nagamine M, Takahara M, Kishibe K, Nagato T, Ishii H, Bandoh N, et al. Sequence variations of Epstein-Barr virus LMP1 gene in nasal NK/T-cell lymphoma. Virus Genes. 2007;**34**(1):47-54

[43] Li Z, Xia Y, Feng LN, Chen JR, Li HM, Cui J, et al. Genetic risk of extranodal natural killer T-cell lymphoma: A genome-wide association study. The Lancet Oncology. 2016;**17**(9):1240-1247

[44] Tamura K, Peterson D, Peterson N, Stecher G, Nei M, Kumar S. MEGA5: Molecular evolutionary genetics analysis using maximum likelihood, evolutionary distance, and maximum parsimony methods. Molecular Biology and Evolution. 2011;**28**(10):2731-2739

[45] Chen JN, Ding YG, Feng ZY, Li HG, He D, Du H, et al. Association of distinctive Epstein-Barr virus variants with gastric carcinoma in Guangzhou, southern China. Journal of Medical Virology. 2010;**82**(4):658-667

[46] Bell MJ, Brennan R, Miles JJ, Moss DJ, Burrows JM, Burrows SR. Widespread sequence variation in Epstein-Barr virus nuclear antigen 1 influences the antiviral T cell response.

The Journal of Infectious Diseases. 2008;**197**(11):1594-1597

[47] Icheva V, Kayser S, Wolff D, Tuve S, Kyzirakos C, Bethge W, et al. Adoptive transfer of epstein-barr virus (EBV) nuclear antigen 1-specific t cells as treatment for EBV reactivation and lymphoproliferative disorders after allogeneic stem-cell transplantation. Journal of Clinical Oncology. 2013;**31**(1):39-48

[48] Fox CP, Haigh TA, Taylor GS, Long HM, Lee SP, Shannon-Lowe C, et al. A novel latent membrane 2 transcript expressed in Epstein-Barr virus-positive NK- and T-cell lymphoproliferative disease encodes a target for cellular immunotherapy. Blood. 2010;**116**(19):3695-3704

[49] Bollard CM, Gottschalk S, Torrano V, Diouf O, Ku S, Hazrat Y, et al. Sustained complete responses in patients with lymphoma receiving autologous cytotoxic T lymphocytes targeting Epstein-Barr virus latent membrane proteins. Journal of Clinical Oncology. 2014;**32**(8):798-808

[50] Cho SG, Kim N, Sohn HJ, Lee SK, Oh ST, Lee HJ, et al. Long-term outcome of Extranodal NK/T cell lymphoma patients treated with postremission therapy using EBV LMP1 and LMP2a-specific CTLs. Molecular Therapy. 2015;**23**(8):1401-1409

[51] Tang KW, Larsson E. Tumour virology in the era of high-throughput genomics. Philosophical Transactions of the Royal Society of London. Series B, Biological Sciences. 2017;**372**(1732):20160265

[52] Plummer M, de Martel C, Vignat J, Ferlay J, Bray F, Franceschi S. Global burden of cancers attributable to infections in 2012: A synthetic analysis. The Lancet Global Health. 2016;**4**(9):e609-e616

[53] Jiang Z, Jhunjhunwala S, Liu J, Haverty PM, Kennemer MI, Guan Y, et al. The effects of hepatitis B virus integration into the genomes of hepatocellular carcinoma patients. Genome Research. 2012;**22**(4):593-601

[54] Hu Z, Zhu D, Wang W, Li W, Jia W, Zeng X, et al. Genome-wide profiling of HPV integration in cervical cancer identifies clustered genomic hot spots and a potential microhomology-mediated integration mechanism. Nature Genetics. 2015;**47**(2):158-163

[55] Liu Y, Lu Z, Xu R, Ke Y. Comprehensive mapping of the human papillomavirus (HPV) DNA integration sites in cervical carcinomas by HPV capture technology. Oncotarget. 2016;7(5):5852-5864

[56] Zapatka M, Borozan I, Brewer DS, Iskar M, Grundhoff A, Alawi M, et al. The landscape of viral associations in human cancers. Nature Genetics. 2020;**52**(3):320-330

[57] Koneva LA, Zhang Y, Virani S, Hall PB, McHugh JB, Chepeha DB, et al. HPV integration in HNSCC correlates with survival outcomes, immune response signatures, and candidate drivers. Molecular Cancer Research. 2018;**16**(1):90-102

[58] Sung WK, Zheng H, Li S, Chen R, Liu X, Li Y, et al. Genome-wide survey of recurrent HBV integration in hepatocellular carcinoma. Nature Genetics. 2012;**44**(7):765-769

[59] Zhao LH, Liu X, Yan HX, Li WY, Zeng X, Yang Y, et al. Genomic and oncogenic preference of HBV integration in hepatocellular carcinoma. Nature Communications. 2016;7:12992

Chapter 2

EBV Genome Mutations and Malignant Proliferations

Sylvie Ranger-Rogez

Abstract

The Epstein-Barr virus (EBV) is a DNA virus with a relatively stable genome. Indeed, genomic variability is reported to be around 0.002%. However, some regions are more variable such as those carrying latency genes and specially *EBNA1*, *-2*, *-LP*, and *LMP1*. Tegument genes, particularly *BNRF1*, *BPLF1*, and *BKRF3*, are also quite mutated. For a long time, it has been considered for this ubiquitous virus, which infects a very large part of the population, that particular strains could be the cause of certain diseases. However, the mutations found, in some cases, are more geographically restricted rather than associated with proliferation. In other cases, they appear to be involved in oncogenesis. The objective of this chapter is to provide an update on changes in viral genome sequences in malignancies associated with EBV. We focused on describing the structure and function of the proteins corresponding to the genes mentioned above in order to understand how certain mutations of these proteins could increase the tumorigenic character of this virus. Mutations described in the literature for these proteins were identified by reporting viral and/or cellular functional changes as they were described.

Keywords: Epstein-Barr virus, lymphoma, carcinoma, mutation, sequence, next generation sequencing

1. Introduction

Epstein-Barr virus (EBV), a ubiquitous gamma-herpesvirus, infects the vast majority of the worldwide human population. This virus was initially discovered in cultured lymphoma cells from patients with Burkitt's lymphoma (BL) in 1964 [1]. During the primary infection, EBV infects epithelial cells of the oropharynx where it actively replicates and also infects B cells where it establishes a life-long latency in the form of an episome located in the host cell nucleus. During latency, EBV may produce nine viral latency proteins, including six so-called "Epstein-Barr Nuclear Antigens" (EBNA1, -2, -3A, -3B, -3C, and -LP), involved in transcriptional regulation, and three "Latent Membrane Proteins" (LMP1, -2A, and -2B), mimicking signals needed for B cell maturation, as well as two small noncoding RNAs (EBER-1 and EBER-2), BamHI-A rightward transcripts (BARTs), and miRNAs. Four different latency programs can be identified, based on the proteins that are expressed (**Table 1**). EBV primary infection, which occurs more often in childhood, is usually asymptomatic in children, whereas it may be responsible for infectious mononucleosis (IM) in teenagers or young adults in western countries. In addition to this nonmalignant disease, EBV can also be associated with diverse malignant pathologies. In particular, EBV is involved in the development of several malignancies of lymphoid

	Program	EBV expressed proteins	Active promoters	B cell type
Latency III	Growth	EBNA-1, -2, -3A, -3B, -3C, -LP	Initially Wp	Naive B cells
		LMP-1, -2A-, -2B	Then Cp	
			LMP promoters	
			EBER-1 and -2p	
Latency II	Default	EBNA-1	Qp, EBER-1 and -2p	
		LMP-1, -2A, -2B	LMP promoters	
Latency I	Latency	EBNA-1	Qp, EBER-1 and -2p	Resting B cells
Latency 0		No protein or LMP-2A	LMP-2Ap	Memory B cells

Table 1.
Proteins expressed during the different latency programs.

origin including endemic Burkitt's lymphoma [2], nasal NK/T lymphoma [3], some Hodgkin's lymphoma [4], and B- or T-cell lymphoproliferations in immunocompromised patients [5]. It is also implicated in epithelial malignancies such as undifferentiated nasopharyngeal carcinoma (NPC) [6] and 10% of cases of gastric carcinoma [7]. Although populations from all geographic areas are infected by the virus, the incidence of the pathologies in which it occurs varies significantly depending on the region [8]. For example, BL occurs mainly in children living in sub-Saharan Africa [9], and the prevalence of NPC is particularly high in adults living in Southern China, Southeast Asia, and Northern Africa [10]. The differences observed in the geographic distribution of these pathologies suggest that there could be various genetic variants of EBV, of different global distributions, and with different levels of transforming capacity. This question of a specific disease variant is raised by many authors and is still being debated. In this chapter, we wish to take inventory of the state of knowledge concerning the variability observed on the most mutated genes among all EBV genes and the possible implications in human pathology.

2. Evolving knowledge of the EBV genome

The fact that the viral genome is relatively large (175 kb), that it is made up of DNA, therefore less variable than if it was an RNA genome, and that it carries repetitive regions, limited its sequencing for a long time. The first published sequences were small fragments of the B95-8 genome; then, the entire B95-8 genome was sequenced in 1984 [11]. The B95-8 strain was the first cultured EBV cell line able to secrete large amounts of viral particles into the culture medium. It was originally obtained from a spontaneous human lymphoblastoid cell line (LCL) established from a North American case of infectious mononucleosis, the 883L cell line, whose virus was used to transform lymphocytes from a cotton top marmoset. Since it was the first strain with a fully published genome, B95-8 has been extensively studied and mapped for transcripts, promoters, and open reading frames.

This first EBV whole genome sequencing was followed by others, and complete viral genome sequences of the cell lines AG876, originating from a Ghanaian case of African BL [12] and GD1, obtained from cord B cells infected with EBV from saliva of an NPC patient in Guangzhou, China [13] were published. Sequences of some genes, mainly latency genes, were also studied, especially in lines established from patients [14, 15]. B95-8, GD1, and AG876 were sequenced by conventional shotgun sequencing (Sanger's method). The comparison of sequences obtained for various cell lines revealed the existence of two types of EBV: type 1 or A, of which B95-8 can be considered as the prototype, and type 2 or B, exemplified by AG876. The main difference between the two types concerns the *EBNA2* gene, with only 70% identity at the nucleotide level and 54% identity in the protein sequence [16]. Additional variations have also been observed in the *EBNA3* genes, but to a lesser extent: 10, 12, and 19% of base pair differences for *EBNA3A, 3B,* and *3C,* respectively [17]. The comparison of viral sequences also highlighted that the B95-8 cell line has a significant 11.8 kb deletion (positions 139,724–151,554) corresponding to some of the *BART* miRNA genes, one of the origins of lytic replication [11], the *LF2* and *LF3* genes, and a part of the *LF1* gene. More complete sequence comprising the B95-8 sequence supplemented with a Raji fragment at the level of deletion has been constructed. It was annotated in 2010 as RefSeq HHV4 (EBV) sequence NC_007605 and is now used as a wild-type strain reference [18].

As adaptation of the virus to *in vitro* culture is possible, thus generating a bias in the results, some authors have preferred to sequence the viral genome directly in samples from patients. Therefore, the sequences GD2, from a Guangzhou NPC biopsy, and HKNPC1, from a Hong Kong NPC biopsy, were published [19, 20], both using a more recent sequencing technique, "next generation sequencing" (NGS). This technology can be used directly on samples or after enrichment, which avoids artifacts due to cellular DNA. Enrichment can be achieved by PCR or cloning into F-factor plasmids, but most frequently, it is carried out using target DNA capture by hybridization. NGS delivers a wealth of information and requires extensive bioinformatic analysis. This technology has made it possible to rapidly increase the number of fully sequenced viral genomes originating from healthy subjects or patients and thus obtain more information.

3. The most variable regions of the genome

Authors who sequenced the entire viral genome and analyzed the genomic variations came to the conclusion that the latent genes harbored the highest numbers of nonsynonymous mutations [20–24]. For example, Liu et al. [25] compared the sequences of nine strains of EBV to GD1, of which they were most closely related, and showed that latency genes were the most mutated. In this study, latent and tegument genes were found to harbor 58.4 to 84.3% of all nonsynonymous mutations detected for each genome. Santpere et al. [26] found that latent genes were twice as mutated as lytic genes. The observation that the latent genes harbor more nucleotide diversity than lytic genes was made regardless of the type of pathology: nasopharyngeal carcinoma [20, 21], NK/T lymphoma [27], endemic Burkitt's lymphoma [22], Hodgkin's lymphoma [22], posttransplant lymphoproliferative disease [22], gastric carcinoma [25], lung carcinoma [23], and also strains originating from infectious mononucleosis [22] or healthy subjects [26]. Why latent genes are the most variable is not clear today. By analyzing their data according to the Yang model [28], Santpere et al. [26] showed that the lytic genes had an evolutionary constraint close to that of the host: a strong purifying selection was objectified for 11 lytic genes. However, signatures of accelerated protein evolution rates were found in coding regions related to virus attachment and entry into host cells. The latency genes, on the other hand, show a positive selection, perhaps in relation to the MHC, which can be the cause of their large diversity. Changes in amino acids (aa) often occur in immune epitopes. Amino acid changes in CD8+ epitopes were described in all latent proteins, while changes in CD4+ epitopes were shown only for EBNA1 and -2 and LMP1 and -2 [20]. However, most codons of the *EBNA3* gene under positive selection are not cytotoxic T-lymphocyte epitopes: either there are epitopes not described to date or the selection relates to other functionalities. The selection of mutants may depend on a difference in immunity in relation to the geography and/or capacity of a strain to infect and persist.

4. Variability of main latency proteins

After the virus enters a host cell, the genome circularizes through recombination of the terminal repeats (TRs) located at each end of the genome to form an episome that will be chromatinized and methylated in the same way as the human genome. Latent transcription programs in B cells are due to the differential activity of epigenetically regulated promoters and take place in three successive waves. The EBNA2 and EBNA-LP, as well as BHRF1, a bcl2 homolog, are the first viral proteins to be expressed, under the dependence of Wp promoter. The two expressed EBNAs and the cellular factor recombination signal-binding protein for immunoglobulin Kappa J region (RBP-Jk) activate then the Cp promoter, which drives the expression of all of the EBNA proteins, while Wp becomes progressively hypermethylated; the transcription will gradually be under Cp control. Subsequently, LMP1, LMP2A, and LMP2B proteins are expressed due to activation of their respective promoters. During latency I or II, Qp promoter controls EBNA1 expression, and Cp methylation is responsible for the five other EBNA silencing. Methylation does not control the Qp promoter, which is switched off by binding to a repressor protein.

As previously developed, latency proteins show the most sequence variations, and among them, EBNA1, EBNA2, EBNA-LP, and LMP1 are the most mutated. The main properties of these proteins are reported in **Table 2**.

4.1 EBNA1

EBNA1, expressed in both latent and lytic EBV infections, was the first EBV protein detected. EBNA1, whose structure (**Figure 1**) and functions have largely been studied [29, 30], is a 641 aa protein. However, EBNA1 proteins frequently exhibit size variations due to differing numbers of gly-ala repeats (aa 89–325). During latency, EBNA1 is the only protein expressed in all forms of latency in proliferating cells and also in all EBV associated malignancies. EBNA1, which acts as a homodimer, is essential for initiating EBV episome replication before mitosis, once per cell cycle, and mitotic segregation of EBV episomes, thus for the maintenance of EBV episome in latently infected cells [31]. The EBNA1 DNA-binding domain is essential but not sufficient for the replication function, and the N-terminal half of EBNA1 is also required. Two EBNA1 regions (aa 8–67 and aa 325–376) are particularly important for this activity, and the point mutations G81 or G425 enhance EBNA1-dependent DNA replication. Inversely, the EBNA1 aa 395–450 region mediates an interaction with the human ubiquitin-specific protease, USP7, which may negatively regulate replication. The partitioning of EBV episomes in two dividing cells requires two viral components: the *ori P* FR element and EBNA1, mainly the central Gly-Arg region aa 325–376 and secondarily the aa 8–67 sequence. EBNA1 also activates the expression of other latency genes participating in immortalization: the regions involved are the central Gly-Arg sequence and the 61–89 region. Interaction with the recognition sites located on FR, DS of *ori P*, and Bam-HI-Q takes place through binding sites located in the C-terminal of EBNA1 (aa 459-607), sequence which also mediates the dimerization of EBNA1 (aa 504–604). Through its interaction with both human casein kinase CK2 (aa 383–395) and cellular ubiquitin-specific protease USP7 (aa 442–448), EBNA1 is also able to disrupt promyelocytic leukemia protein

DOI: http://dx.doi.org/10.5772/intechopen.93194

Protein	Role/ localization	Main properties
EBNA1	Latency	Initiation of viral episome replication before mitosis
		Mitotic segregation of EBV episomes
		Transcription of other latency genes (Cp and LMPp enhancer)
		Degradation of promyelocytic leukemia protein (PML) bodies
		Cellular transcription regulation
EBNA2	Latency	Viral and cellular transcription factor
		Initiation and maintenance of B cell immortalization
		Blocking of methylation sites for BZLF-1 binding
EBNA-LP	Latency	Coactivator of the transcriptional activator EBNA2
LMP1	Latency	Similarity to constitutively activated CD40
		Constitutive activation of cell pathways
		Maintenance of EBV latency and control of cell migration
BNRF1	Tegument	Establishment of latency and cell immortalization
		Increase in the number of cellular centrioles
BPLF1	Tegument	Downregulation of viral ribonucleotide reductase (RR)
		Disruption of damaged DNA repair
		Decreasing of innate immunity
BKRF3	Tegument	DNA replication and repair—viral DNA mutagenesis prevention

Table 2.
Main properties of proteins developed in this chapter.

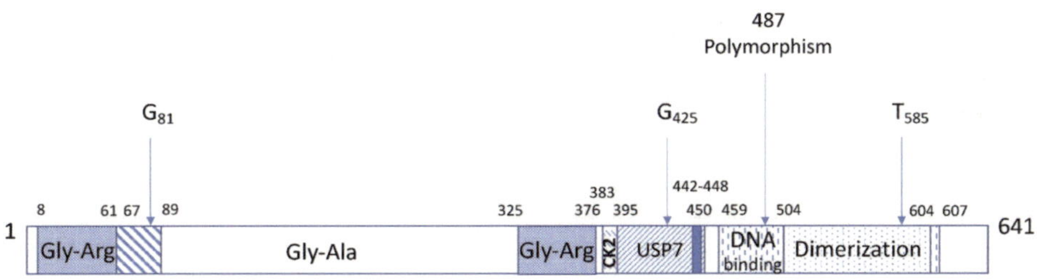

Figure 1.
Schematic representation of basic structure of EBNA1 protein with the different motifs and their position. Gly-Arg: region rich in Gly-Arg; Gly-Ala: Gly-Ala repeats; CK2: interaction with human casein kinase, CK2; USP7: interaction with the human ubiquitin specific protease, USP7; DNA binding: DNA-binding domain; Dimerization: region that mediates the dimerization of EBNA1. The different mutations discussed are noted.

(PML) bodies and degrade PML. In addition to its role in latent infection, EBNA1 can therefore participate in lytic infection by overcoming suppression by PML proteins [32]. Indeed, PML proteins and nuclear bodies were found to suppress lytic infection by EBV. Recently [33], organization in an oligomeric hexameric ring form was described for the EBNA1 DNA-binding domain, the oligomeric interface pivoting around residue T585. Mutations occurring on this residue had both positive and negative effects on EBNA1-dependent DNA replication and episome maintenance.

Based on polymorphisms observed at 15 codons, Bhatia et al. [14] reported two strains named P (prototype) and V (variant), each having two subtypes defined by the aa at position 487 (P-ala, P-thr, V-pro, and V-leu). They detected mostly the P-thr and the V-leu variants, respectively, in African and American BL tumors, but these findings were not confirmed by another group who reported different spectra of EBNA1 subtypes according to different geographical areas in both healthy patients and BL tumors [34]. A fifth subtype, V-val, was later recognized in South-East Asia and was found to be prevalent in NPC samples by numerous authors [20, 35–37]. These findings suggest that the V-val variant might adapt particularly well to the nasopharyngeal epithelium or that this strain possesses an increased oncogenic potential. Indeed, most of the variant codons, localized in the DNA-binding domain, may have an impact on the EBV phenotype resulting in impaired ability to transform B-lymphocytes [30]. However, other reports observed that this subtype had no tumor-specific expression [38], and it is likely that it probably represents a dominant EBNA1 subtype in Asian regions, not found in other areas of the world [8, 23, 25]. The P-thr subtype is the most commonly observed in peripheral blood of American and African subjects as well as in African tumors. In our experience, P-thr is also the most prevalent in France and particularly in the course of lymphoproliferative diseases.

Apart from these mutations, others have been reported. For example, Borozan et al. [39] looked at gastric carcinomas and mainly found two mutations already described in NPC, H418L and A439T, located outside the DNA-binding domain and common in both NPC and GC but uncommon in other EBV isolates, from lymphomas or healthy subjects. They also described a new mutation, T85A, positioned in the region required for transcriptional activation of other latency genes and thus able to modify this function. Wang et al. [23] described the substitution T585I. T585 is subject to substitutions, and T585 polymorphism is found frequently in NPC tumors and Burkitt's lymphoma. T585I was previously found, and this strain was defective in replication and maintenance of the viral episome [40], as well as deficient in suppressing lytic cycle gene transcription and lytic DNA replication.

In summary, EBNA1 V-val variant seems to be a geographic variant almost exclusively present in South-East Asia. Conversely, mutations T85 and T585, which occur in functional regions of the protein, could have biological consequences and especially the substitution T585I, which promotes lytic replication and is found in NPC.

4.2 EBNA2

EBNA2, a 487 aa protein, is expressed *in vivo* during latency III shortly after infection of B cells or in lymphomas occurring in immunocompromised patients and in LCL. As mentioned above, the variations in EBNA2 make it possible to classify EBV as types 1 and 2 (or A and B) since only 70% identity at the nucleotide level and 54% homology in the protein sequence were observed. The overall structure of the EBNA2 protein (**Figure 2**) is characterized by poly-P and poly-RG areas, this last one being a protein-protein and protein-nucleic acid interaction domain important for efficient cell growth transformation, and nine regions conserved throughout the gene [41]. EBNA2 acts principally as a transcription factor and contains three categories of domains critical for its transcription regulation function: transactivation domains (TAD), self-association domains (SAD), and nuclear localization signals (NLS). EBNA2 does not bind directly to DNA. It uses cell proteins as adapters to access viral or cellular enhancer and promoter sites. The C-terminal TAD (aa 448–471) is able to recruit components of basic transcriptional machinery as well as chromatin modifiers and can bind to the viral coactivator EBNA-LP, while the N-terminal TAD (aa 1–58) cannot bind EBNA-LP, although its activity can be enhanced by this protein. Two SADs (aa 1–58 and 97–121), separated by the polyproline stretch, were identified in the N-terminal region [42]. An additional third one has been reported, localized in a nonconserved region, and flanked by the second SAD and the adapter region [43]. EBNA2 contributes to B-cell immortalization, and

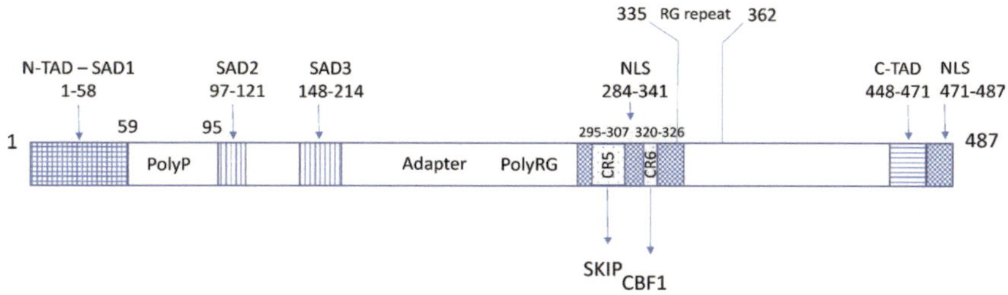

Figure 2.
Schematic representation of basic structure of EBNA2 protein with the different motifs and their position. The two transactivation domains (TADs), the three self-association domains (SADs), and the two nuclear localization signals (NLSs) are mentioned. Poly P: area rich in P; PolyRG: area rich in RG; CR5: conserved region 5, which interacts with SKIP (Ski-interacting protein); CR6: conserved region 6, which interacts with CBF1 (C promoter-binding factor 1). The different mutations discussed are mentioned.

it has been demonstrated that type 1 EBV, which is predominantly found in EBV-associated diseases, immortalizes B cells *in vitro* much more efficiently than type 2 [44], which is predominantly determined by sequence variation in the C-terminus of EBNA2 [45]. During the early events of EBV infection in resting B cells, EBNA2 initiates the transcription of a cascade of primary and secondary viral and cellular target genes and therefore is responsible for the initiation of immortalization by reprogramming the resting state into a proliferative state. For this, EBNA2 interacts with chromatin remodelers and as a transcription factor cofactor [46]. Mühe et al. [47] demonstrated that the first 150 N-terminal aa of EBNA2 are important for the initiation of immortalization. EBNA2 is also involved in immortalization maintenance; the region implicated here (aa 295–378) includes the conserved regions CR5 (aa 295–307) and CR6 (aa 320–326), particularly important for this function. CR5 mediates the contact between EBNA2 and SKIP (Ski-interacting protein), and CR6 is the CBF1 (C promoter-binding factor 1) or RBP-Jk targeting domain. Mechanisms to initiate and maintain B cell immortalization are not completely understood today.

Wang et al. [41], working on 25 EBV-associated GCs, 56 NPCs, and 32 throat washings from healthy donors in Northern China, described 4 EBNA2 subtypes according to the presence of a deletion, namely subtypes E2-A (no aa deletion), E2-B (aa 294Q deletion), E2-C (aa 357K and 358G deletion), and E2-D (aa 357K, 358G, and 294Q deletion). The E2-A subtype exhibited six nonsilent mutations, P291T, R413G, I438L, E476G, P484H, and I486T; the substitution P291T was present in six NPC E2-D and six NPC E2-C. The substitution R413G was detected in E2-C for one patient. They found that E2-A and E2-C were dominant in the samples they analyzed and that the E2-D pattern was detected only in the NPC specimens. The mutation R163M was detected in all samples. This mutation has previously been described worldwide and in different diseases.

Mutations 357 and 358 occurred in the RG domain (aa 335–362), a downregulator of EBNA2 activation of the LMP1 promoter [48]. Moreover, aa 357–363 (KGKSRDK) constitutes the PKC phosphorylation site, which can reduce the amounts of EBNA2/CBF1 complex formed. EBNA2 is suspected to be involved in the development of malignancies as a result of sequence variations most frequently affecting its regulation function.

Interestingly, *EBNA2* entire-gene deletion has been shown in some endemic BL cell lines such as P3HR1, Daudi, Sav, Oku, and Ava [49]; it remains to determine if this deletion occurs classically *in vivo* in African BL.

In short, geographic variants were not formally demonstrated for EBNA2. Among the described mutations, the most interesting are those occurring in the PKC phosphorylation site because they can activate the Cp and/or LMP1p and thus increase the production of latency proteins.

4.3 EBNA-LP (EBNA-leader protein)

EBNA-LP, like EBNA2 and concomitantly with EBNA2, is expressed shortly after the infection of B cells in healthy individuals as well as in EBV-related malignant diseases in immunodeficient patients and LCLs. EBNA-LP acts mostly as a coactivator of the transcriptional activator EBNA2, thus inducing the expression of some cellular genes, including *cyclin D2* [50], or viral genes, that is, *LMP1* [51], *LMP2b,* and *Cp* and therefore having an important role in B cell immortalization. EBNA-LP also can directly interact with several cell proteins such as tumor suppressors or proteins involved in apoptosis or cell cycle regulation.

EBNA-LP is comprised of a variable number of 66 aa repetitive units, corresponding to the variable number of W1 and W2 exons located in the EBV internal repeat IR1, followed by a unique 45 aa domain, encoded by two unique 3′ exons Y1 and Y2 (**Figure 3**). Therefore, EBNA-LP protein may vary in size according to the number of W1–W2 repeats contained in each EBV isolate. By convention, the protein annotation is based on a single W repeat isoform (**Figure 4**). In this configuration, the protein has 110 aa. Conserved regions were identified in the N extremity of the protein (CR1 to CR3, respectively, aa 11–33, 45–52, and 55–62, implicated in EBNA2 binding), and in the C-terminal region (CR4 and CR5, respectively, aa 76–82 and 101–110). CR3 and a serine within W2 (S35) were demonstrated to be important for EBNA2 coactivation. EBV-mediated B cell immortalization maps to the W1W2 repeated domains and requires at least two IR1 repetitions to be effective, but a number greater than or equal to 5 is optimal [53]. Some interactions with cell proteins are mediated by the repeated W1W2 N-terminus [54]. *EBNA-LP* gene transcription initiates from the W promoter (Wp) residing in each IR1 repeat during the early stages of infection, and multiple EBNA-LP protein isoforms are produced. During the later stages of infection and in LCLs, transcription initiates from the C promotor (Cp) [55]. The level of transcription initiated by Cp compared to Wp varies according to different circumstances [56].

About 15% of BL tumors host a virus, which uses exclusively the W promoter, expressing an EBV atypical latency program [49], harboring EBNA1, EBNA3A, 3B, 3C, and a truncated form of EBNA-LP. In these cases, EBV genome lacks the *EBNA2*

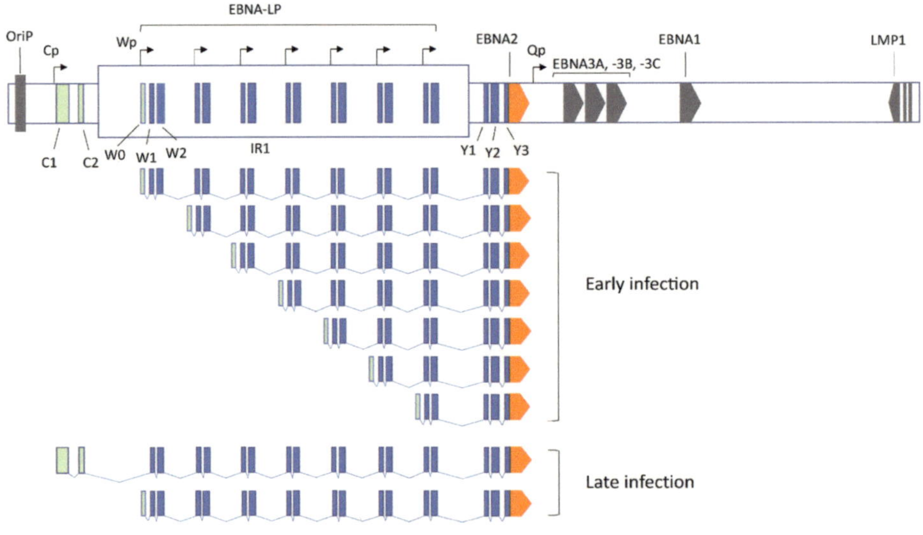

Figure 3.
Schematic representation of the IR1 region of EBV genome (according to Ref. [52]). The promoters Wp, Cp, and Qp are represented, as well as the different proteins expressed according to the stage of infection.

gene and the unique Y1Y2 exons of *EBNA-LP*. This was firstly described in P3HR1 and Daudi BL cell lines [57]. Subsequently, these cells were shown to be more resistant to apoptosis than cells infected by wild-type virus, what would be related to the truncated shape of EBNA-LP.

Given the difficulty of sequencing repetitive regions, only few authors have sequenced the IR1 region, including the EBNA-LP coding region. Previous studies identified two EBNA-LP distinct isoforms, type 1 and type 2 variants, based on the presence of G8/T12 or V8/A12 in exon W1 [58]. The Q54R substitution was also described in exon W2 from an African type 2 spontaneous lymphoblastoid cell line LCL [59]. Despite this, a high degree of conservation was reported for the Wp promoter and the W1-W2 intron, while the most diversity was observed for the BWRF1 ORF, which only shows 80% homology between various strains, and for Y exons [60]. The sequence variations in the Y exons, and especially the Y2 exon, made it possible to define four main subgroups, called A, B, C, and Z. The Akata strain belongs to subgroup A and B95-8 to subgroup B. Subgroup Z is found in type 2 EBVs, and the C subtype is characterized by V95E and V102I. Finally, it has been reported that tumor-derived strains are more prone to interstrain genetic exchange in IR1 [60].

4.4 LMP1

LMP1 is considered to be the main oncogenic protein in EBV. LMP1 is a multifunctional self-aggregating protein essential for the transformation of human B cells and rodent fibroblasts [61]. It is a 386 aa protein comprising a 24 aa cytosolic N-terminal (NT) segment, a 162 aa portion consisting of six transmembrane (TM) domains, and a 200 aa cytosolic C-terminal (CT) domain (**Figure 5**) [62]. The NT domain plays an important role in the orientation and anchoring of LMP1 to the membrane and its constitutive aggregation, thus contributing to the transforming function of LMP1 [63]. The TM region is involved in the localization of LMP1 at the level of lipid rafts in the membrane, thus inducing its clustering to activate signaling from the CT tail. It is remarkable that the $F_{38}LWY_{41}$ pattern in the first

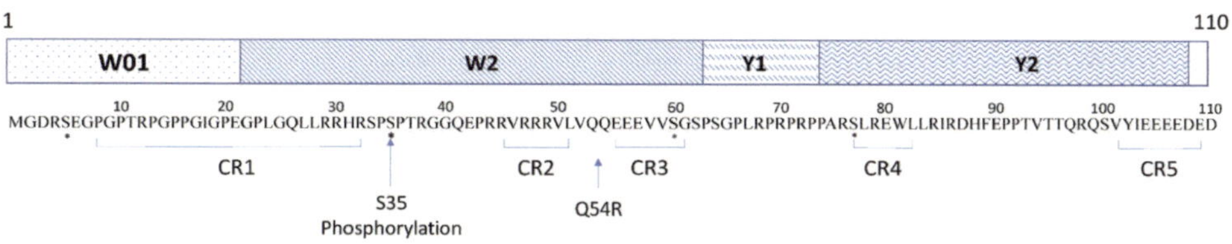

Figure 4.
Sequence of EBNA-LP protein, with the position of the corresponding exons opposite. Conserved regions are represented as well as the key positions. Phosphorylated serins are mentioned by an asterisk.

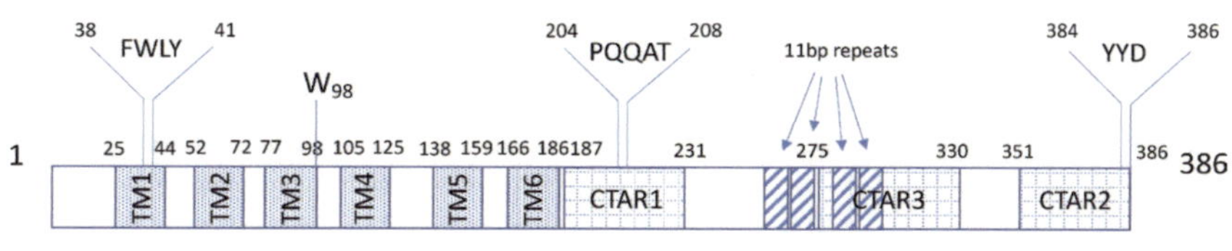

Figure 5.
Schematic representation of basic structure of LMP1 protein with the different motifs and their position. TM1–6: transmembrane domains 1–6. The FWLY pattern in TM1 and W98 in TM3 are essential for the association of TM1–2 with TM3–6 and oligomerization signaling. CTAR1–3: carboxyl-terminal activating regions 1–3. PQQAT pattern is necessary for the attachment of TRAF adapters. YYD pattern is essential for binding the TNF receptor-associated death domain (TRADD) adapter.

transmembrane fragment (TM1) and a second pattern consisting of aa W98 in TM3 are essential for the association of TM domains (1–2) with TM domains (3–6) as well as for the oligomerization and signaling of LMP1 [64]. The CT part is involved in the activation of LMP1-induced cell signaling pathways, including two important regions, CTAR1/TES1 and CTAR2/TES2 (Carboxyl-Terminal Activating Region/Transformation Effector Site) critical for EBV-mediated B-cell growth transformation [65]. Together, these regions mimic CD40, a member of the tumor necrosis factor (TNF) receptor family and key B-cell costimulatory receptor, thus enabling the recruitment of cell adapters associated with the TNF receptor family, TNF receptor-associated factors (TRAFs). The CTAR1 region includes the P_{204}-X-Q_{206}-X-T_{208} consensus pattern necessary for the attachment of TRAF adapters, specifically TRAF1, TRAF2, TRAF3, and TRAF5 [66]. Within the CTAR2 region, the Y_{384}-Y_{385}-D_{386} pattern is essential for binding the TNF receptor-associated death domain (TRADD) adapter. There is a third region, CTAR3 (aa 232–350), that is not essential for *in vitro* B cell immortalization and is less well known [67]. In this region located between CTAR1 and CTAR2 (aa 253–302), a variable number of repeat 11 aa elements (4 repeats for B95-8) exist.

LMP1 acts principally as a viral pseudoreceptor, which regulates host cell signal transduction by constitutive activation of cell pathways as mitogen-activated protein kinase (MAPK) pathways and principally the extracellular regulated kinases 1 and 2 (ERK1/2), c-Jun amino-terminal kinases 1–3 (JNK1–3), and p38 isoform pathways. LMP1 also induces the phosphatidylinositol 3-kinase (PI3K) pathway, which contributes to survival signals [68] and transcription of activator protein 1 (AP1) [69], PI3K, and AP1 pathways, therefore playing a major role in proliferation and cell cycle control. LMP1 is also responsible for the activation of JAK/STAT and interferon regulatory factor 7 (IRF7) pathways and for aberrant constitutive NF-kB activation. Indeed, the CTAR1 PXQXT pattern is able to engage TRAFs, leading finally to the activation of noncanonical NF-kB pathway that controls processing of the NF-kB2/p100 precursor [70]. The CTAR2 YYD pattern is in turn implicated in the activation of the canonical NF-kB pathway [71] after binding of tumor necrosis factor receptor superfamily member 1A (TNFRSF1A)-associated via TRADD and receptor-inter-acting protein 1 (RIP1). A wider region of LMP1 seems to be responsible for binding RIP1 (aa 351–386), compared to TRADD (aa 375–386) [72]. NF-kB is considered to be the principal factor by which LMP1 regulates gene expression and modifies cell behavior [73]. Activation of NF-kB is associated with upregulation of anti-apoptotic genes [32, 74] and downregulation of pro-apoptotic factors, as well as induction of tumorigenesis-associated B-cell activation markers [75, 76]. CTAR3, less well defined, seems to activate SUMOylation pathways and participate in the maintenance of EBV latency and control of cell migration, a hallmark of oncogenesis [77, 78].

Besides its ability to transform B cells, during the latency state, LMP1 seems also to be able to facilitate the release of virions from B cells during lytic replication [32].

Variations in the LMP1 sequence have been widely studied, particularly in the context of its impact on clinical occurrence or evolution. A 30 bp deletion (del30), resulting in a 10 aa loss in the C-terminal (aa 343–352), was first described in the Cao cell isolate from a Chinese NPC [79]. In addition, this isolate harbored numerous substitutions. A high prevalence of the same deletion, as reviewed by Chang et al. [8], was found in Asian NPC biopsy tissues [80, 81], in lymphomas and EBV-related gastric cancers from Eastern Asia [82] and in Asian nasal NK/T-cell lymphomas [83, 84]. Del30 was shown to be often associated with the G335D mutation in NPC, and such strains were reported to have a greater transforming activity *in vitro* than the reference LMP1 [85, 86]. If the 30 bp deletion is partly localized to CTAR2, it does not alter NF-kB activation [87] and finally does not modify signaling

Region concerned	Mutation	Consequence of the mutation	References
	E2D		[86, 96]
	H3R/L		[86, 94, 96]
NT domain (24AA)	R7S		[86]
Mutation rate: 0.33	G8A		[86]
	G11S		[86]
	P12A		[86]
	R13P		[86, 97]
	R17L		[86, 97]
	L25I/M		[13, 86, 94, 96, 97]
TM1 (20AA)	L33I		[86]
Mutation rate: 0.15	V43I		[86, 94, 96]
	D46N		[86, 94, 96, 97]
	S57A		[86, 94, 96]
TM2 (21AA)	M61I		[86, 97]
Mutation rate: 0.19	163 V/L		[86, 94, 96, 97]
	F70V		[94]
	A82G		[86, 97]
	C84G/V		[86, 97]
TM3 (21AA)	I85L	Homo-oligomerization and/or interaction with other molecules in lipid rafts	[86, 94, 96, 97]
Mutation rate: 0.24	T91I		[86]
	I95S		[86]
	F106V/Y		[86, 96, 97]
	V110L		[86]
	F112Y	Homo-oligomerization and/or interaction with other molecules in lipid rafts	[94]
TM4 (21AA)	G115A		[86]
Mutation rate: 0.43	W116C		[86]
	V119A		[86]
	L120F		[86]
	I122L		[86, 94, 97]
	I124G/V	I124V + I152L: increased NF-κB activation *in vitro*	[94, 96, 98]
	L126F		[86, 94, 96, 97]
	M129I	M129I: increased LMPI half-life in epithelial cells	[86, 94, 96, 97]
	II37L		[86]
TM5 (22AA) Mutation rate: 0.18	F144I/D	FI44I/D + D150A/L + L151I: increased NF-κB activation *in vitro*	[86, 97, 98]
	D150A	D150A/L + F144I/D + LI511: increased NF-κB activation *in vitro*	[86, 97, 98]
	L151I	L151I + F144I/D + D150A/L: increased NF-κB activation *in vitro*	[86, 94, 96–98]
	II52L	I152L + I124V: increased NF-κB activation *in vitro*	[94, 96, 98]

Region concerned	Mutation	Consequence of the mutation	References
TM6 (21AA)	L178M		[86, 97]
Mutation rate: 0.05	Q189P		[86, 97]
	S192T		[86, 97]
	G212ST	G212S: Erk activation, thus c-Fos induction and binding to API site	[86, 94, 96, 97]
		SNQ pattern (212–214) + del30 in NK/T biopsies	
CTAR1 (45AA)	H213N	SNQ pattern (212–214) + del30 in NK/T biopsies	[86, 94, 96]
Mutation rate: 0.15	E214Q	SNQ pattern (212–214) + del30 in NK/T biopsies	[86, 94, 96]
	H225L		[86]
	S229T		[94, 97]
	P245H		[86]
	G248D		[86]
	D250N		[86]
	N251Y		[23]
	G252D		[86]
	del275–279		[86, 94]
	D293G		[94]
	D298A		[86, 97]
	H308Q		[23]
CTAR3 (56AA)	S309N	S309N + del30 + del15 in NK/T biopsies	[86, 94, 96, 97]
Mutation rate: 0.20	G318K		[97]
	Q322E/N/K		[86, 94, 96, 97]
	E325D		[97]
	V327L		[97]
	E328A		[86]
	K330A		
	G331Q		[86]
	D333N		[94]
	Q334R		[86, 94, 96, 97]
	G335D/S		[86, 94, 97]
	del343–352		[23, 86, 94, 97]
	L338S/P		[86, 94, 96, 97]
	G344A/D		[86, 97]
	H348D		[94]
CTAR2 (35AA)	H352R		[94, 97]
Mutation rate: 0.23	G353D		[86]
	G355A/I		[86]
	D356M		[86]
	S366T/Q/A	S366T: Erk activation, thus c-Fos induction and binding to API site	[86, 94, 96, 97]

Table 3.
LMP1 mutations described in the literature.

properties [88]. However, it is clear that strains bearing del30 are selected over the wt-LMP1 variants in NK/T-cell lymphomas [83] and NPC tumors [89]. Given that del30 strains have been currently detected in normal carriers [90] or in various EBV-associated diseases [91], and, because of a low prevalence of del30 strains in samples from Africa, North America, and Europe [8, 92], it is generally admitted that LMP1 del30 may represent a geographic polymorphism rather than a disease-associated polymorphism [93]. In a study, we carried out in France in patients with NK/T lymphoma, we found a del30 EBV in 4/4 biopsies studied and in 46.1% of total blood samples analyzed, while in a control population, the deletion was present in 4.8% of cases [94]. Other deletions were also described, such as the rare C terminal 69 bp deletion reported to weakly activate the AP1 transcription factor [95], or the 15 bp deletion (aa 275–279) frequently encountered in Western Europe [94].

Otherwise, numerous substitutions have been described in LMP1 (**Table 3**), particularly in the N-terminal extremity. Some authors have made attempts to classify viral strains by taking into account these substitutions with the aim of highlighting a viral implication in certain pathologies [99]. Thus, Mainou and Raab-Traub [88] classified EBV into seven variants, namely Alaskan, China 1, China 2, Med+, Med-, NC, and B95-8, all having the same *in vitro* transforming potential and signaling properties. Zuercher et al. [98] mentioned two polymorphisms, I124V/I152L and F144I/D150A/L151I, which seem to be markers of increased NF-kB activation *in vitro*. Lei et al. [96] distinguished four models according to the substitutions occurring in both the *LMP1* gene and its promoter. The patients suffering from NPC that they studied all carried a strain belonging to pattern B, while the BLs were distributed among the four patterns. Many authors recognize two evolutionarily distinct clusters, Asian-derived EBV strains including GD2, HKNPC1, and Akata strains and non-Asian and African/American strains including AG876, B95-8, and Mutu strains, suggesting that the *LMP1* gene could be used as a geographic marker [25, 97].

Finally, it should be noted that LMP1 carries a molecular signature of accelerated evolution rate probably due to positive selection as deduced from a significant proportion of nonsignificant variations [26].

So, regarding LMP1, which is the most oncogenic latency protein, two geographic clusters appear to exist corresponding to an Asian variant and a non-Asiatic variant. The described 30 bp deletion is mainly present on Asian strains, and it shows an obvious tropism for nasopharynx. Although many substitutions have been described, little work is done to analyze changes in LMP1 properties based on these substitutions. NPC could be associated with a particular strain, but this remains to be confirmed.

5. Variability of tegument proteins

After the latency proteins, the tegument proteins carry the most changes, and among them, the most mutated are BNRF1, BPLF1, and BKRF3, which will be detailed, as well as BBRF2. This latter protein appears to play an important role in viral infectivity [100], but its structure and function are poorly known today. For this reason, BBRF2 will not be developed here.

5.1 BNRF1

EBV major tegument protein BNRF1 contains 1318 aa, and its structure is shown schematically in **Figure 6**. BNRF1 is a member of a protein family with homology to the cellular purine biosynthesis enzyme FGARAT. BNRF1 is involved in the

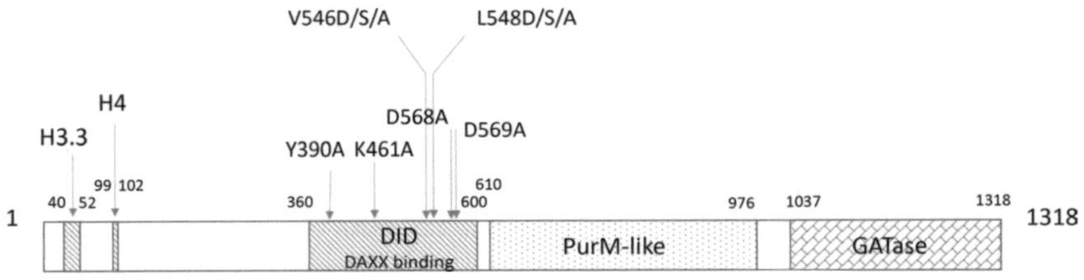

Figure 6.
Schematic representation of basic structure of BNRF1 protein with the different motifs and their position. H3.3 and H4 regions, respectively, involved in binding to H3.3 and H4. DID: DAXX-interaction domain, domain implicated in binding to DAXX (death-domain associated protein-6) histone chaperone. PurM-like domain and GATase domain were noted, as well as the different mutations discussed.

establishment of latency and cell immortalization by hijacking the antiviral DAXX (death domain-associated protein-6) histone chaperone [101]. BNRF1 seems to have lost conventional purine biosynthesis activity. It forms a stable quaternary complex with DAXX histone-binding domain (HBD), H3.3 and H4 [102], responsible for BNRF1 localization to PML nuclear bodies involved in antiviral intrinsic resistance and transcriptional repression of host cells. In the presence of BNFR1, DAXX can no longer collaborate with ATRX to assemble histone variant H3.3 into repressive chromatin at GC-rich repetitive DNA. Binding to DAXX, histone H3.3 and histone H4 occur, respectively, via the BNRF1 DAXX interaction domain (DID) (aa 360–600) and BNRF1 residues 40–52 and 99–102. Huang et al. [102] demonstrated that the quaternary complex formation is abrogated when dual mutations V546D/L548D and D568A/D569A occurred on BRNF1 DID and is partially diminished *in vitro* in case of dual mutations Y390A/K461A and V546S/L548S on BNRF1 DID. BNRF1 mutations at K461A, Y390A/K461A, V546S/L548S or Y390A, V546A/L548A, and D568A/569A moderately or severely reduced BNRF1 colocalization at PML nuclear bodies, respectively. A PurM-like domain (610–976) and a GATase domain (1037–1318) were defined. It has also recently been shown that BNRF1 can cause an abnormal increase in the number of cellular centrioles [103]. This phenomenon can lead to aneuploidy or structural chromosome abnormalities and, possibly, to carcinogenesis. The gene regions concerned have not been described.

BNRF1 is reported to be one of the most frequently mutated tegument proteins. It is interesting to note that a nonsense mutation was described in C666–1, an EBV-positive NPC cell line, with no major structural alterations in the BNRF1-deleted virus [92].

So, the mutations described for BNRF1 do not appear to correspond to a particular geographical distribution. On the other hand, some mutations seem to be able to modify DNA chromatinization, thus affecting the transcription, and therefore have important consequences on cell functioning.

5.2 BPLF1

BPLF1, the largest EBV protein (3149 aa), is a late lytic tegument protein. BPLF1 possesses a deubiquitinating (DUB) activity. BPLF1 is able to downregulate viral ribonucleotide reductase (RR) activity, by deubiquitination of the large subunit RR1 [104], and to specifically deubiquitinate proliferating cell nuclear antigen (PCNA), a DNA polymerase processivity factor, thus disrupting the repair of damaged DNA [105]. By triggering activation of repair pathways and co-opting DNA repair and replication factors, the virus could create genomic instability. The DUB activity is carried by the first 246 aa of the N-terminal region, and the C61 residue of the catalytic triad (Cys-His-Asp) is essential for activity [104]. BPLF1 relocalizes

Pol η to nuclear sites of viral DNA production, thereby bypassing DNA damage [106]. This mechanism contributes to efficient production of infectious virus.

BPLF1 is also able to deubiquinate cell factors, such as TRAF6, NEMO, and IkBα, leading to TLR signaling inhibition through both MyD88- and TRIF-dependent pathways, thus decreasing innate immune responses by reduced NF-kB activation and proinflammatory cytokine production [107]. It is noteworthy that the same catalytic active site also carries a deneddylating activity shown to target cullin ring ligases, potentially affecting viral replication and infectivity [108]. The role of BPLF1 to help drive human B-cell immortalization and lymphoma formation has also been discussed [109].

Sequencing of various viral strains has shown that BPLF1 is one of the proteins with the greatest number of changes [20, 24, 110]. Most of these mutations are not analyzed in detail, but Kwok et al. [21], working on the sequences of eight NPC biopsy specimens, reported two nonsynonymous mutations in the N-terminal region of the protein that exhibit deubiquitinating activity. The same finding was reported by Simbiri et al. [110], who also described 3 C-terminal mutations (L2935P, P2987L, and R3005Q). A single-nucleotide deletion coupled with a single-nucleotide insertion three nucleotides away was reported by Zeng et al. [13] in a NPC strain. As a result, two aa substitutions (GA/EG) were predicted to occur. Tu et al. [24] undertook phylogenetic analysis based on several reported EBV genome sequences and some major genes as *BPLF1*. They observed that EBV Asian subtypes clustered as a separate branch from the non-Asian ones.

So, as with other proteins, it seems that the Asian strains carry a protein different from the other strains. Substitutions occurring in the region carrying the deubiquitinase activity could have biological consequences.

5.3 BKRF3

BKRF3 is a small protein (255 aa), which belongs to the early lytic gene family, and encodes an uracil-DNA glycosylase (UDG), which removes inappropriate uracil residues from DNA. BKRF3 excises uracil bases incorporated in double-stranded DNA due to uracil misincorporation or more often cytosine deamination [111, 112]. BKRF3 participates in DNA replication and repair and prevents viral DNA mutagenesis. BKRF3 shares substantial similarity in overall structure with the one UDG family. Four of the five catalytic motifs are completely conserved (aa 90–94, 110–114, 146–149, 191–192), whereas the fifth domain (aa 213–229) carries a seven-residue insertion in the leucine loop [113]. In addition, the 29 N-terminal aa carry a nuclear localization signal (sequence KRKQ). Only changes in BKRF3 that do not severely affect viral replication can be retained, but it may be considered that these mutations cause a change in virus-cell interrelations.

6. Conclusion

The aim of this chapter was to take stock of the most frequently observed variations in the EBV genome and more particularly to see if some of these variations are considered to be involved in tumor pathology. The candidate viral genes concerned are numerous; those developed here are the most affected, and the mutations reported in the literature have been identified. Some mutations have been well studied, in particular as regards their impact on the structure or functionality of the protein or the cellular consequences of these modifications. However, most mutations have only been described. If a tumorigenic impact of viral mutations is not yet certain, many authors agree that geographic variants exist, and it seems clear that Asian strains have different characteristics from non-Asian strains. Further work is necessary to complete the mass of information and analysis, not at the level of one or several genes, but at the level of the entire genome.

Author details

Sylvie Ranger-Rogez[1,2]

1 Department of Virology, University Hospital Dupuytren, Limoges, France

2 Faculty of Pharmacy, UMR CNRS 7276, UMR INSERM 1262, Limoges, France

*Address all correspondence to: sylvie.rogez@unilim.fr

References

[1] Epstein MA, Achong BG, Barr YM. Virus particles in cultured lymphoblasts from BURKITT'S lymphoma. Lancet. 1964;**1**:702-703. DOI: 10.1016/s0140-6736(64)91524-7

[2] Zur Hausen H, Schulte-Holthausen H. Presence of EB virus nucleic acid homology in a "virus-free" line of Burkitt tumour cells. Nature. 1970;**227**:245-248. DOI: 10.1038/227245a0

[3] Aozasa K, Takakuwa T, Hongyo T, Yang W-I. Nasal NK/T-cell lymphoma: Epidemiology and pathogenesis. International Journal of Hematology. 2008;**87**:110-117. DOI: 10.1007/s12185-008-0021-7

[4] Weiss LM, Movahed LA, Warnke RA, Sklar J. Detection of Epstein-Barr viral genomes in reed-Sternberg cells of Hodgkin's disease. The New England Journal of Medicine. 1989;**320**:502-506. DOI: 10.1056/NEJM198902233200806

[5] Hamilton-Dutoit SJ, Pallesen G, Franzmann MB, Karkov J, Black F, Skinhøj P, et al. AIDS-related lymphoma. Histopathology, immunophenotype, and association with Epstein-Barr virus as demonstrated by in situ nucleic acid hybridization. The American Journal of Pathology. 1991;**138**:149-163

[6] Klein G, Giovanella BC, Lindahl T, Fialkow PJ, Singh S, Stehlin JS. Direct evidence for the presence of Epstein-Barr virus DNA and nuclear antigen in malignant epithelial cells from patients with poorly differentiated carcinoma of the nasopharynx. Proceedings of the National Academy of Sciences of the United States of America. 1974;**71**:4737-4741. DOI: 10.1073/pnas.71.12.4737

[7] Imai S, Koizumi S, Sugiura M, Tokunaga M, Uemura Y, Yamamoto N, et al. Gastric carcinoma: Monoclonal epithelial malignant cells expressing Epstein-Barr virus latent infection protein. Proceedings of the National Academy of Sciences of the United States of America. 1994;**91**:9131-9135. DOI: 10.1073/pnas.91.19.9131

[8] Chang CM, Yu KJ, Mbulaiteye SM, Hildesheim A, Bhatia K. The extent of genetic diversity of Epstein-Barr virus and its geographic and disease patterns: A need for reappraisal. Virus Research. 2009;**143**:209-221. DOI: 10.1016/j.virusres.2009.07.005

[9] Ziegler JL. Burkitt's lymphoma. The New England Journal of Medicine. 1981;**305**:735-745. DOI: 10.1056/NEJM198109243051305

[10] Kumar S, Mahanta J. Aetiology of nasopharyngeal carcinoma. A review. Indian Journal of Cancer. 1998;**35**:47-56

[11] Baer R, Bankier AT, Biggin MD, Deininger PL, Farrell PJ, Gibson TJ, et al. DNA sequence and expression of the B95-8 Epstein-Barr virus genome. Nature. 1984;**310**:207-211. DOI: 10.1038/310207a0

[12] Dolan A, Addison C, Gatherer D, Davison AJ, McGeoch DJ. The genome of Epstein-Barr virus type 2 strain AG876. Virology. 2006;**350**:164-170. DOI: 10.1016/j.virol.2006.01.015

[13] Zeng M-S, Li D-J, Liu Q-L, Song L-B, Li M-Z, Zhang R-H, et al. Genomic sequence analysis of Epstein-Barr virus strain GD1 from a nasopharyngeal carcinoma patient. Journal of Virology. 2005;**79**:15323-15330. DOI: 10.1128/JVI.79.24.15323-15330.2005

[14] Bhatia K, Raj A, Guitierrez MI, Judde JG, Spangler G, Venkatesh H, et al. Variation in the sequence of Epstein Barr virus nuclear antigen 1 in normal peripheral blood lymphocytes

and in Burkitt's lymphomas. Oncogene. 1996;**13**:177-181

[15] Walling DM, Shebib N, Weaver SC, Nichols CM, Flaitz CM, Webster-Cyriaque J. The molecular epidemiology and evolution of Epstein-Barr virus: Sequence variation and genetic recombination in the latent membrane protein-1 gene. The Journal of Infectious Diseases. 1999;**179**:763-774. DOI: 10.1086/314672

[16] Adldinger HK, Delius H, Freese UK, Clarke J, Bornkamm GW. A putative transforming gene of Jijoye virus differs from that of Epstein-Barr virus prototypes. Virology. 1985;**141**:221-234. DOI: 10.1016/0042-6822(85)90253-3

[17] Sample J, Young L, Martin B, Chatman T, Kieff E, Rickinson A, et al. Epstein-Barr virus types 1 and 2 differ in their EBNA-3A, EBNA-3B, and EBNA-3C genes. Journal of Virology. 1990;**64**:4084-4092

[18] Tzellos S, Farrell PJ. Epstein-Barr virus sequence variation-biology and disease. Pathogens. 2012;**1**:156-174. DOI: 10.3390/pathogens1020156

[19] Liu P, Fang X, Feng Z, Guo Y-M, Peng R-J, Liu T, et al. Direct sequencing and characterization of a clinical isolate of Epstein-Barr virus from nasopharyngeal carcinoma tissue by using next-generation sequencing technology. Journal of Virology. 2011;**85**:11291-11299. DOI: 10.1128/JVI.00823-11

[20] Kwok H, Tong AHY, Lin CH, Lok S, Farrell PJ, Kwong DLW, et al. Genomic sequencing and comparative analysis of Epstein-Barr virus genome isolated from primary nasopharyngeal carcinoma biopsy. PLoS One. 2012;7:e36939. DOI: 10.1371/journal.pone.0036939

[21] Kwok H, Wu CW, Palser AL, Kellam P, Sham PC, Kwong DLW, et al. Genomic diversity of Epstein-Barr virus genomes isolated from primary nasopharyngeal carcinoma biopsy samples. Journal of Virology. 2014;**88**:10662-10672. DOI: 10.1128/JVI.01665-14

[22] Palser AL, Grayson NE, White RE, Corton C, Correia S, Ba Abdullah MM, et al. Genome diversity of Epstein-Barr virus from multiple tumor types and normal infection. Journal of Virology. 2015;**89**:5222-5237. DOI: 10.1128/JVI.03614-14

[23] Wang S, Xiong H, Yan S, Wu N, Lu Z. Identification and characterization of Epstein-Barr virus genomes in lung carcinoma biopsy samples by next-generation sequencing technology. Scientific Reports. 2016;**6**:26156. DOI: 10.1038/srep26156

[24] Tu C, Zeng Z, Qi P, Li X, Guo C, Xiong F, et al. Identification of genomic alterations in nasopharyngeal carcinoma and nasopharyngeal carcinoma-derived Epstein-Barr virus by whole-genome sequencing. Carcinogenesis. 2018;**39**:1517-1528. DOI: 10.1093/carcin/bgy108

[25] Liu Y, Yang W, Pan Y, Ji J, Lu Z, Ke Y. Genome-wide analysis of Epstein-Barr virus (EBV) isolated from EBV-associated gastric carcinoma (EBVaGC). Oncotarget. 2016;7:4903-4914. DOI: 10.18632/oncotarget.6751

[26] Santpere G, Darre F, Blanco S, Alcami A, Villoslada P, Mar Albà M, et al. Genome-wide analysis of wild-type Epstein-Barr virus genomes derived from healthy individuals of the 1,000 genomes project. Genome Biology and Evolution. 2014;**6**:846-860. DOI: 10.1093/gbe/evu054

[27] Peng R-J, Han B-W, Cai Q-Q, Zuo X-Y, Xia T, Chen J-R, et al. Genomic and transcriptomic landscapes of Epstein-Barr virus in extranodal natural killer T-cell lymphoma. Leukemia.

2019;**33**:1451-1462. DOI: 10.1038/s41375-018-0324-5

[28] Yang Z. PAML 4: Phylogenetic analysis by maximum likelihood. Molecular Biology and Evolution. 2007;**24**:1586-1591. DOI: 10.1093/molbev/msm088

[29] Frappier L. The Epstein-Barr virus EBNA1 protein. Scientifica (Cairo). 2012;**2012**:438204. DOI: 10.6064/2012/438204

[30] Wilson JB, Manet E, Gruffat H, Busson P, Blondel M, Fahraeus R. EBNA1: Oncogenic activity, immune evasion and biochemical functions provide targets for novel therapeutic strategies against Epstein-Barr virus-associated cancers. Cancers (Basel). 2018;**10**(4):109. 130 pages. DOI: 10.3390/cancers10040109

[31] Young LS, Murray PG. Epstein-Barr virus and oncogenesis: From latent genes to tumours. Oncogene. 2003;**22**:5108-5121. DOI: 10.1038/sj.onc.1206556

[32] Li C-W, Jheng B-R, Chen B-S. Investigating genetic-and-epigenetic networks, and the cellular mechanisms occurring in Epstein-Barr virus-infected human B lymphocytes via big data mining and genome-wide two-sided NGS data identification. PLoS One. 2018;**13**:e0202537. DOI: 10.1371/journal.pone.0202537

[33] Deakyne JS, Malecka KA, Messick TE, Lieberman PM. Structural and functional basis for an EBNA1 hexameric ring in Epstein-Barr virus episome maintenance. Journal of Virology. 2017;**91**(19):17 pages. DOI: 10.1128/JVI.01046-17

[34] Habeshaw G, Yao QY, Bell AI, Morton D, Rickinson AB. Epstein-Barr virus nuclear antigen 1 sequences in endemic and sporadic Burkitt's lymphoma reflect virus strains prevalent in different geographic areas. Journal of Virology. 1999;**73**:965-975

[35] Gutiérrez MI, Raj A, Spangler G, Sharma A, Hussain A, Judde JG, et al. Sequence variations in EBNA-1 may dictate restriction of tissue distribution of Epstein-Barr virus in normal and tumour cells. The Journal of General Virology. 1997;**78**(Pt 7):1663-1670. DOI: 10.1099/0022-1317-78-7-1663

[36] Wang W-Y, Chien Y-C, Jan J-S, Chueh C-M, Lin J-C. Consistent sequence variation of Epstein-Barr virus nuclear antigen 1 in primary tumor and peripheral blood cells of patients with nasopharyngeal carcinoma. Clinical Cancer Research. 2002;**8**:2586-2590

[37] Zhang X-S, Wang H-H, Hu L-F, Li A, Zhang R-H, Mai H-Q, et al. V-val subtype of Epstein-Barr virus nuclear antigen 1 preferentially exists in biopsies of nasopharyngeal carcinoma. Cancer Letters. 2004;**211**:11-18. DOI: 10.1016/j.canlet.2004.01.035

[38] Sandvej K, Zhou XG, Hamilton-Dutoit S. EBNA-1 sequence variation in Danish and Chinese EBV-associated tumours: Evidence for geographical polymorphism but not for tumour-specific subtype restriction. The Journal of Pathology. 2000;**191**:127-131. DOI: 10.1002/(SICI)1096-9896(200006)191:2<127::AID-PATH614>3.0.CO;2-E

[39] Borozan I, Zapatka M, Frappier L, Ferretti V. Analysis of Epstein-Barr virus genomes and expression profiles in gastric adenocarcinoma. Journal of Virology. 2018;**92**(2):17 pages. DOI: 10.1128/JVI.01239-17

[40] Dheekollu J, Malecka K, Wiedmer A, Delecluse H-J, Chiang AKS, Altieri DC, et al. Carcinoma-risk variant of EBNA1 deregulates Epstein-Barr virus episomal latency. Oncotarget. 2017;**8**:7248-7264. DOI: 10.18632/oncotarget.14540

[41] Wang X, Wang Y, Wu G, Chao Y, Sun Z, Luo B. Sequence analysis of Epstein-Barr virus EBNA-2 gene coding amino acid 148-487 in nasopharyngeal and gastric carcinomas. Virology Journal. 2012;**9**:49. DOI: 10.1186/1743-422X-9-49

[42] Friberg A, Thumann S, Hennig J, Zou P, Nössner E, Ling PD, et al. The EBNA-2 N-terminal transactivation domain folds into a dimeric structure required for target gene activation. PLoS Pathogens. 2015;**11**:e1004910. DOI: 10.1371/journal.ppat.1004910

[43] Tsui S, Schubach WH. Epstein-Barr virus nuclear protein 2A forms oligomers in vitro and in vivo through a region required for B-cell transformation. Journal of Virology. 1994;**68**:4287-4294

[44] Rickinson AB, Young LS, Rowe M. Influence of the Epstein-Barr virus nuclear antigen EBNA 2 on the growth phenotype of virus-transformed B cells. Journal of Virology. 1987;**61**:1310-1317

[45] Tzellos S, Correia PB, Karstegl CE, Cancian L, Cano-Flanagan J, McClellan MJ, et al. A single amino acid in EBNA-2 determines superior B lymphoblastoid cell line growth maintenance by Epstein-Barr virus type 1 EBNA-2. Journal of Virology. 2014;**88**:8743-8753. DOI: 10.1128/JVI.01000-14

[46] Lu F, Chen H-S, Kossenkov AV, DeWispeleare K, Won K-J, Lieberman PM. EBNA2 drives formation of new chromosome binding sites and target genes for B-cell master regulatory transcription factors RBP-jκ and EBF1. PLoS Pathogens. 2016;**12**:e1005339. DOI: 10.1371/journal.ppat.1005339

[47] Mühe J, Wang F. Species-specific functions of Epstein-Barr virus nuclear antigen 2 (EBNA2) reveal dual roles for initiation and maintenance of B cell immortalization. PLoS Pathogens. 2017;**13**:e1006772. DOI: 10.1371/journal.ppat.1006772

[48] Peng C-W, Zhao B, Kieff E. Four EBNA2 domains are important for EBNALP coactivation. Journal of Virology. 2004;**78**:11439-11442. DOI: 10.1128/JVI.78.20.11439-11442.2004

[49] Kelly G, Bell A, Rickinson A. Epstein-Barr virus-associated Burkitt lymphomagenesis selects for downregulation of the nuclear antigen EBNA2. Nature Medicine. 2002;**8**:1098-1104. DOI: 10.1038/nm758

[50] Sinclair AJ, Palmero I, Peters G, Farrell PJ. EBNA-2 and EBNA-LP cooperate to cause G0 to G1 transition during immortalization of resting human B lymphocytes by Epstein-Barr virus. The EMBO Journal. 1994;**13**:3321-3328

[51] Nitsche F, Bell A, Rickinson A. Epstein-Barr virus leader protein enhances EBNA-2-mediated transactivation of latent membrane protein 1 expression: A role for the W1W2 repeat domain. Journal of Virology. 1997;**71**:6619-6628

[52] Kempkes B et al. EBNA2 and its coactivator EBNA-LP. In: Münz C, editor. Epstein Barr Virus. Vol. 2. One Herpes Virus: Many Diseases. Heidelberg, Berlin: Springer; 2015. pp. 35-59

[53] Tierney RJ, Kao K-Y, Nagra JK, Rickinson AB. Epstein-Barr virus BamHI W repeat number limits EBNA2/EBNA-LP coexpression in newly infected B cells and the efficiency of B-cell transformation: A rationale for the multiple W repeats in wild-type virus strains. Journal of Virology. 2011;**85**:12362-12375. DOI: 10.1128/JVI.06059-11

[54] Han I, Harada S, Weaver D, Xue Y, Lane W, Orstavik S, et al. EBNA-LP associates with cellular proteins

including DNA-PK and HA95. Journal of Virology. 2001;**75**:2475-2481. DOI: 10.1128/JVI.75.5.2475-2481.2001

[55] Woisetschlaeger M, Yandava CN, Furmanski LA, Strominger JL, Speck SH. Promoter switching in Epstein-Barr virus during the initial stages of infection of B lymphocytes. Proceedings of the National Academy of Sciences of the United States of America. 1990;**87**:1725-1729. DOI: 10.1073/pnas.87.5.1725

[56] Chelouah S, Cochet E, Couvé S, Balkaran S, Robert A, May E, et al. New interactors of the truncated EBNA-LP protein identified by mass spectrometry in P3HR1 Burkitt's lymphoma cells. Cancers (Basel). 2018;**10**(12):14 pages. DOI: 10.3390/cancers10010012

[57] Jones MD, Foster L, Sheedy T, Griffin BE. The EB virus genome in Daudi Burkitt's lymphoma cells has a deletion similar to that observed in a non-transforming strain (P3HR-1) of the virus. The EMBO Journal. 1984;**3**:813-821

[58] Finke J, Rowe M, Kallin B, Ernberg I, Rosén A, Dillner J, et al. Monoclonal and polyclonal antibodies against Epstein-Barr virus nuclear antigen 5 (EBNA-5) detect multiple protein species in Burkitt's lymphoma and lymphoblastoid cell lines. Journal of Virology. 1987;**61**:3870-3878

[59] McCann EM, Kelly GL, Rickinson AB, Bell AI. Genetic analysis of the Epstein-Barr virus-coded leader protein EBNA-LP as a co-activator of EBNA2 function. The Journal of General Virology. 2001;**82**:3067-3079. DOI: 10.1099/0022-1317-82-12-3067

[60] Ba Abdullah MM, Palermo RD, Palser AL, Grayson NE, Kellam P, Correia S, et al. Heterogeneity of the Epstein-Barr virus (EBV) major internal repeat reveals evolutionary mechanisms of EBV and a functional defect in the prototype EBV strain B95-8. Journal of Virology. 2017;**91**(23):25 pages. DOI: 10.1128/JVI.00920-17

[61] Dirmeier U, Neuhierl B, Kilger E, Reisbach G, Sandberg ML, Hammerschmidt W. Latent membrane protein 1 is critical for efficient growth transformation of human B cells by Epstein-Barr virus. Cancer Research. 2003;**63**:2982-2989

[62] Li H-P, Chang Y-S. Epstein-Barr virus latent membrane protein 1: Structure and functions. Journal of Biomedical Science. 2003;**10**:490-504. DOI: 10.1007/bf02256110

[63] Coffin WF, Erickson KD, Hoedt-Miller M, Martin JM. The cytoplasmic amino-terminus of the latent membrane protein-1 of Epstein-Barr virus: Relationship between transmembrane orientation and effector functions of the carboxy-terminus and transmembrane domain. Oncogene. 2001;**20**:5313-5330. DOI: 10.1038/sj.onc.1204689

[64] Yasui T, Luftig M, Soni V, Kieff E. Latent infection membrane protein transmembrane FWLY is critical for intermolecular interaction, raft localization, and signaling. Proceedings of the National Academy of Sciences of the United States of America. 2004;**101**:278-283. DOI: 10.1073/pnas.2237224100

[65] Izumi KM, Kaye KM, Kieff ED. The Epstein-Barr virus LMP1 amino acid sequence that engages tumor necrosis factor receptor associated factors is critical for primary B lymphocyte growth transformation. Proceedings of the National Academy of Sciences of the United States of America. 1997;**94**:1447-1452. DOI: 10.1073/pnas.94.4.1447

[66] Mosialos G, Birkenbach M, Yalamanchili R, VanArsdale T, Ware C, KieffE.TheEpstein-Barrvirustransforming

protein LMP1 engages signaling proteins for the tumor necrosis factor receptor family. Cell. 1995;**80**:389-399. DOI: 10.1016/0092-8674(95)90489-1

[67] Izumi KM, Cahir McFarland ED, Riley EA, Rizzo D, Chen Y, Kieff E. The residues between the two transformation effector sites of Epstein-Barr virus latent membrane protein 1 are not critical for B-lymphocyte growth transformation. Journal of Virology. 1999;**73**:9908-9916

[68] Dawson CW, Tramountanis G, Eliopoulos AG, Young LS. Epstein-Barr virus latent membrane protein 1 (LMP1) activates the phosphatidylinositol 3-kinase/Akt pathway to promote cell survival and induce actin filament remodeling. The Journal of Biological Chemistry. 2003;**278**:3694-3704. DOI: 10.1074/jbc.M209840200

[69] Kieser A, Kilger E, Gires O, Ueffing M, Kolch W, Hammerschmidt W. Epstein-Barr virus latent membrane protein-1 triggers AP-1 activity via the c-Jun N-terminal kinase cascade. The EMBO Journal. 1997;**16**:6478-6485. DOI: 10.1093/emboj/16.21.6478

[70] Lavorgna A, Harhaj EW. EBV LMP1: New and shared pathways to NF-κB activation. Proceedings of the National Academy of Sciences of the United States of America. 2012;**109**:2188-2189. DOI: 10.1073/pnas.1121357109

[71] Gewurz BE, Mar JC, Padi M, Zhao B, Shinners NP, Takasaki K, et al. Canonical NF-kappaB activation is essential for Epstein-Barr virus latent membrane protein 1 TES2/CTAR2 gene regulation. Journal of Virology. 2011;**85**:6764-6773. DOI: 10.1128/JVI.00422-11

[72] Eliopoulos AG, Young LS. LMP1 structure and signal transduction. Seminars in Cancer Biology. 2001;**11**:435-444. DOI: 10.1006/scbi.2001.0410

[73] Cahir-McFarland ED, Carter K, Rosenwald A, Giltnane JM, Henrickson SE, Staudt LM, et al. Role of NF-kappa B in cell survival and transcription of latent membrane protein 1-expressing or Epstein-Barr virus latency III-infected cells. Journal of Virology. 2004;**78**:4108-4119. DOI: 10.1128/jvi.78.8.4108-4119.2004

[74] Calderwood MA, Venkatesan K, Xing L, Chase MR, Vazquez A, Holthaus AM, et al. Epstein-Barr virus and virus human protein interaction maps. Proceedings of the National Academy of Sciences of the United States of America. 2007;**104**:7606-7611. DOI: 10.1073/pnas.0702332104

[75] Liu M-T, Chang Y-T, Chen S-C, Chuang Y-C, Chen Y-R, Lin C-S, et al. Epstein-Barr virus latent membrane protein 1 represses p53-mediated DNA repair and transcriptional activity. Oncogene. 2005;**24**:2635-2646. DOI: 10.1038/sj.onc.1208319

[76] Wang LW, Jiang S, Gewurz BE. Epstein-Barr virus LMP1-mediated oncogenicity. Journal of Virology. 2017;**91**(21):11 pages. DOI: 10.1128/JVI.01718-16

[77] Bentz GL, Whitehurst CB, Pagano JS. Epstein-Barr virus latent membrane protein 1 (LMP1) C-terminal-activating region 3 contributes to LMP1-mediated cellular migration via its interaction with Ubc9. Journal of Virology. 2011;**85**:10144-10153. DOI: 10.1128/JVI.05035-11

[78] Bentz GL, Moss CR, Whitehurst CB, Moody CA, Pagano JS. LMP1-induced sumoylation influences the maintenance of Epstein-Barr virus latency through KAP1. Journal of Virology. 2015;**89**:7465-7477. DOI: 10.1128/JVI.00711-15

[79] Hu LF, Zabarovsky ER, Chen F, Cao SL, Ernberg I, Klein G, et al. Isolation and sequencing of the

Epstein-Barr virus BNLF-1 gene (LMP1) from a Chinese nasopharyngeal carcinoma. The Journal of General Virology. 1991;**72**(Pt 10):2399-2409. DOI: 10.1099/0022-1317-72-10-2399

[80] Miller WE, Edwards RH, Walling DM, Raab-Traub N. Sequence variation in the Epstein-Barr virus latent membrane protein 1. The Journal of General Virology. 1994;**75**(Pt 10):2729-2740. DOI: 10.1099/0022-1317-75-10-2729

[81] Tan E-L, Peh S-C, Sam C-K. Analyses of Epstein-Barr virus latent membrane protein-1 in Malaysian nasopharyngeal carcinoma: High prevalence of 30-bp deletion, Xho1 polymorphism and evidence of dual infections. Journal of Medical Virology. 2003;**69**:251-257. DOI: 10.1002/jmv.10282

[82] Mori S, Itoh T, Tokunaga M, Eizuru Y. Deletions and single-base mutations within the carboxy-terminal region of the latent membrane protein 1 oncogene in Epstein-Barr virus-related gastric cancers of southern Japan. Journal of Medical Virology. 1999;**57**:152-158. DOI: 10.1002/(sici)1096-9071(199902)57:2<152::aid-jmv11>3.0.co;2-k

[83] Chiang AK, Wong KY, Liang AC, Srivastava G. Comparative analysis of Epstein-Barr virus gene polymorphisms in nasal T/NK-cell lymphomas and normal nasal tissues: Implications on virus strain selection in malignancy. International Journal of Cancer. 1999;**80**:356-364. DOI: 10.1002/(sici)1097-0215(19990129)80:3<356::aid-ijc4>3.0.co;2-d

[84] Nagamine M, Takahara M, Kishibe K, Nagato T, Ishii H, Bandoh N, et al. Sequence variations of Epstein-Barr virus LMP1 gene in nasal NK/T-cell lymphoma. Virus Genes. 2007;**34**:47-54. DOI: 10.1007/s11262-006-0008-5

[85] Li SN, Chang YS, Liu ST. Effect of a 10-amino acid deletion on the oncogenic activity of latent membrane protein 1 of Epstein-Barr virus. Oncogene. 1996;**12**:2129-2135

[86] Cheung ST, Leung SF, Lo KW, Chiu KW, Tam JS, Fok TF, et al. Specific latent membrane protein 1 gene sequences in type 1 and type 2 Epstein-Barr virus from nasopharyngeal carcinoma in Hong Kong. International Journal of Cancer. 1998;**76**:399-406. DOI: 10.1002/(sici)1097-0215(19980504)76:3<399::aid-ijc18>3.0.co;2-6

[87] Farrell PJ. Signal transduction from the Epstein-Barr virus LMP-1 transforming protein. Trends in Microbiology. 1998;**6**:175-177; discussion 177-178. DOI: 10.1016/s0966-842x(98)01262-1

[88] Mainou BA, Raab-Traub N. LMP1 strain variants: Biological and molecular properties. Journal of Virology. 2006;**80**:6458-6468. DOI: 10.1128/JVI.00135-06

[89] Edwards RH, Sitki-Green D, Moore DT, Raab-Traub N. Potential selection of LMP1 variants in nasopharyngeal carcinoma. Journal of Virology. 2004;**78**:868-881. DOI: 10.1128/jvi.78.2.868-881.2004

[90] Correa RM, Fellner MD, Alonio LV, Durand K, Teyssié AR, Picconi MA. Epstein-Barr virus (EBV) in healthy carriers: Distribution of genotypes and 30 bp deletion in latent membrane protein-1 (LMP-1) oncogene. Journal of Medical Virology. 2004;**73**:583-588. DOI: 10.1002/jmv.20129

[91] Lorenzetti MA, Gantuz M, Altcheh J, De Matteo E, Chabay PA, Preciado MV. Distinctive Epstein-Barr virus variants associated with benign and malignant pediatric pathologies: LMP1 sequence characterization and linkage with other viral gene

polymorphisms. Journal of Clinical Microbiology. 2012;**50**:609-618. DOI: 10.1128/JCM.05778-11

[92] Tso KK-Y, Yip KY-L, Mak CK-Y, Chung GT-Y, Lee S-D, Cheung S-T, et al. Complete genomic sequence of Epstein-Barr virus in nasopharyngeal carcinoma cell line C666-1. Infectious Agents and Cancer. 2013;**8**:29. DOI: 10.1186/1750-9378-8-29

[93] Zhang X-S, Song K-H, Mai H-Q, Jia W-H, Feng B-J, Xia J-C, et al. The 30-bp deletion variant: A polymorphism of latent membrane protein 1 prevalent in endemic and non-endemic areas of nasopharyngeal carcinomas in China. Cancer Letters. 2002;**176**:65-73. DOI: 10.1016/s0304-3835(01)00733-9

[94] Halabi MA, Jaccard A, Moulinas R, Bahri R, Al Mouhammad H, Mammari N, et al. Clonal deleted latent membrane protein 1 variants of Epstein-Barr virus are predominant in European extranodal NK/T lymphomas and disappear during successful treatment. International Journal of Cancer. 2016;**139**:793-802. DOI: 10.1002/ijc.30128

[95] Larcher C, Bernhard D, Schaadt E, Adler B, Ausserlechner MJ, Mitterer M, et al. Functional analysis of the mutated Epstein-Barr virus oncoprotein LMP1(69del): Implications for a new role of naturally occurring LMP1 variants. Haematologica. 2003;**88**:1324-1335

[96] Lei H, Li T, Li B, Tsai S, Biggar RJ, Nkrumah F, et al. Epstein-Barr virus from Burkitt lymphoma biopsies from Africa and South America share novel LMP-1 promoter and gene variations. Scientific Reports. 2015;**5**:16706. DOI: 10.1038/srep16706

[97] Lin Z, Wang X, Strong MJ, Concha M, Baddoo M, Xu G, et al. Whole-genome sequencing of the Akata and Mutu Epstein-Barr virus strains. Journal of Virology. 2013;**87**:1172-1182. DOI: 10.1128/JVI.02517-12

[98] Zuercher E, Butticaz C, Wyniger J, Martinez R, Battegay M, Boffi El Amari E, et al. Genetic diversity of EBV-encoded LMP1 in the Swiss HIV Cohort Study and implication for NF-κb activation. PLoS One. 2012;**7**:e32168. DOI: 10.1371/journal.pone.0032168

[99] Edwards RH, Seillier-Moiseiwitsch F, Raab-Traub N. Signature amino acid changes in latent membrane protein 1 distinguish Epstein-Barr virus strains. Virology. 1999;**261**:79-95. DOI: 10.1006/viro.1999.9855

[100] Masud HMAA, Yanagi Y, Watanabe T, Sato Y, Kimura H, Murata T. Epstein-Barr virus BBRF2 is required for maximum infectivity. Microorganisms. 2019;**7**(705):14 pages. DOI: 10.3390/microorganisms7120705

[101] Tsai K, Thikmyanova N, Wojcechowskyj JA, Delecluse H-J, Lieberman PM. EBV tegument protein BNRF1 disrupts DAXX-ATRX to activate viral early gene transcription. PLoS Pathogens. 2011;**7**:e1002376. DOI: 10.1371/journal.ppat.1002376

[102] Huang H, Deng Z, Vladimirova O, Wiedmer A, Lu F, Lieberman PM, et al. Structural basis underlying viral hijacking of a histone chaperone complex. Nature Communications. 2016;**7**:12707. DOI: 10.1038/ncomms12707

[103] Shumilov A, Tsai M-H, Schlosser YT, Kratz A-S, Bernhardt K, Fink S, et al. Epstein-Barr virus particles induce centrosome amplification and chromosomal instability. Nature Communications. 2017;**8**:14257. DOI: 10.1038/ncomms14257

[104] Whitehurst CB, Ning S, Bentz GL, Dufour F, Gershburg E, Shackelford J, et al. The Epstein-Barr virus (EBV)

deubiquitinating enzyme BPLF1 reduces EBV ribonucleotide reductase activity. Journal of Virology. 2009;**83**:4345-4353. DOI: 10.1128/JVI.02195-08

[105] Whitehurst CB, Vaziri C, Shackelford J, Pagano JS. Epstein-Barr virus BPLF1 deubiquitinates PCNA and attenuates polymerase η recruitment to DNA damage sites. Journal of Virology. 2012;**86**:8097-8106. DOI: 10.1128/JVI.00588-12

[106] Dyson OF, Pagano JS, Whitehurst CB. The translesion polymerase pol η is required for efficient Epstein-Barr virus infectivity and is regulated by the viral deubiquitinating enzyme BPLF1. Journal of Virology. 2017;**91**(19):14 pages. DOI: 10.1128/JVI.00600-17

[107] van Gent M, Braem SGE, de Jong A, Delagic N, Peeters JGC, Boer IGJ, et al. Epstein-Barr virus large tegument protein BPLF1 contributes to innate immune evasion through interference with toll-like receptor signaling. PLoS Pathogens. 2014;**10**:e1003960. DOI: 10.1371/journal.ppat.1003960

[108] Gastaldello S, Hildebrand S, Faridani O, Callegari S, Palmkvist M, Guglielmo CD, et al. A deneddylase encoded by Epstein–Barr virus promotes viral DNA replication by regulating the activity of cullin-RING ligases. Nature Cell Biology. 2010;**12**:351-361. DOI: 10.1038/ncb2035

[109] Whitehurst CB, Li G, Montgomery SA, Montgomery ND, Su L, Pagano JS. Knockout of Epstein-Barr virus BPLF1 retards B-cell transformation and lymphoma formation in humanized mice. MBio. 2015;**6**(5):11 pages. DOI: 10.1128/mBio.01574-15

[110] Simbiri KO, Smith NA, Otieno R, Wohlford EEM, Daud II, Odada SP, et al. Epstein-Barr virus genetic variation in lymphoblastoid cell lines derived from Kenyan pediatric population. PLoS One. 2015;**10**:e0125420. DOI: 10.1371/journal.pone.0125420

[111] Lu C-C, Huang H-T, Wang J-T, Slupphaug G, Li T-K, Wu M-C, et al. Characterization of the uracil-DNA glycosylase activity of Epstein-Barr virus BKRF3 and its role in lytic viral DNA replication. Journal of Virology. 2007;**81**:1195-1208. DOI: 10.1128/JVI.01518-06

[112] Su M-T, Liu I-H, Wu C-W, Chang S-M, Tsai C-H, Yang P-W, et al. Uracil DNA glycosylase BKRF3 contributes to Epstein-Barr virus DNA replication through physical interactions with proteins in viral DNA replication complex. Journal of Virology. 2014;**88**:8883-8899. DOI: 10.1128/JVI.00950-14

[113] Géoui T, Buisson M, Tarbouriech N, Burmeister WP. New insights on the role of the gamma-herpesvirus uracil-DNA glycosylase leucine loop revealed by the structure of the Epstein-Barr virus enzyme in complex with an inhibitor protein. Journal of Molecular Biology. 2007;**366**:117-131. DOI: 10.1016/j.jmb.2006.11.007

Chapter 3

Epstein-Barr Virus-Associated Gastric Cancer: Old Entity with New Relevance

Hugo Manuel Lopes de Sousa, Joana Patrícia Costa Ribeiro and Mafalda Basílio Timóteo

Abstract

Gastric cancer (GC) represents a major public health issue worldwide, being the fifth most common cancer and one of the leading causes of death by cancer. In 2014, The Cancer Genome Atlas (TCGA) established that tumors positive for Epstein-Barr virus (EBV) are considered a specific subtype of GC (EBVaGC). Several meta-analyses have shown that EBVaGC represents almost 10% of all gastric cancer worldwide, with small differences in the geographic distribution. This tumor subtype has a high potential of being clinically relevant and studies have shown that it has specific features, a better prognosis, and increased overall survival. In this review, we summarize some of the most frequent aspects of EBVaGC, including the specific features of this GC subtype, data regarding the potential steps of EBVaGC carcinogenesis, and perspectives on treatment opportunities.

Keywords: Epstein-Barr virus, gastric cancer, carcinogenesis, p53, PDL-1, immunotherapy

1. Introduction (Gastric cancer)

1.1 Epidemiology

Gastric cancer (GC) represents a major public health issue worldwide, being the fifth most common cancer and one of the leading causes of death by cancer [1, 2]. GC affects more than 1,000,000 people per year and leads to approximately 783,000 deaths each year, corresponding to 5.7% of new cases and 8.2% of all cancer related deaths (**Figure 1**). Worldwide, GC incidence has a distinct geographic distribution pattern [3, 4] (**Figure 2**). The highest incidence rates are registered in Eastern Asia and Central/Eastern Europe, while Northern America and Africa have the lowest incidence rates [1, 5].

There seems to exist some ethnic/racial disparities in the distribution of GC [6, 7]; nevertheless studies showed that this may be the explained by the different expositions to GC risk factors such as dietary, salt intake, and *Helicobacter pylori* infection [5, 8]. Furthermore, despite the worrying high mortality associated with GC, the incidence of GC globally has been declining since 1990. This trend is mostly due to the falling rates of non-cardia GC, which is explained by the improvement of hygienic conditions and early detection of cancer strategies [9].

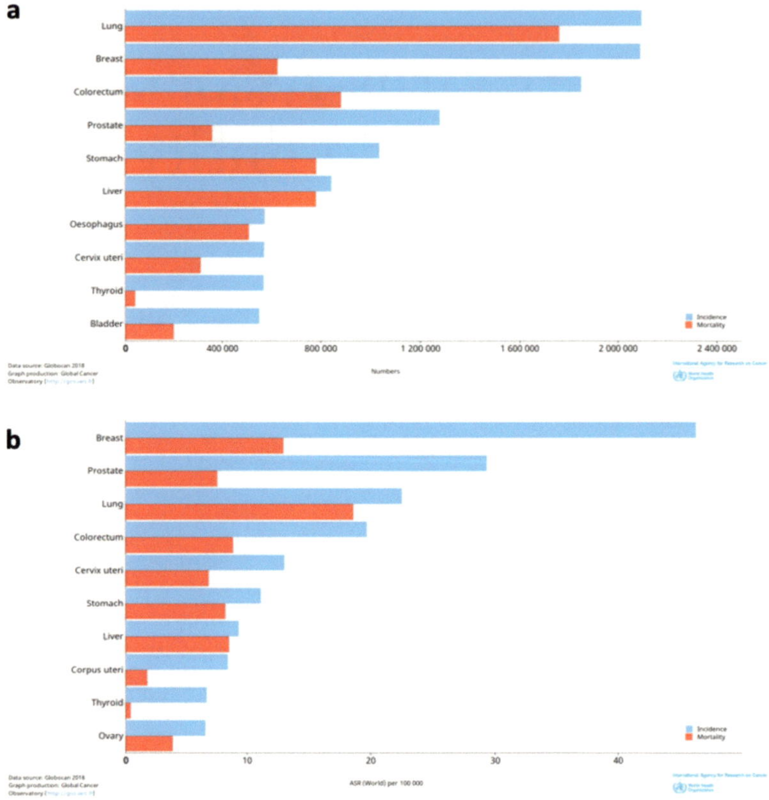

Figure 1.
Estimated number (a) of incident cases and deaths of gastric cancer worldwide and (b) age-standardized rates (GLOBOCAN 2018).

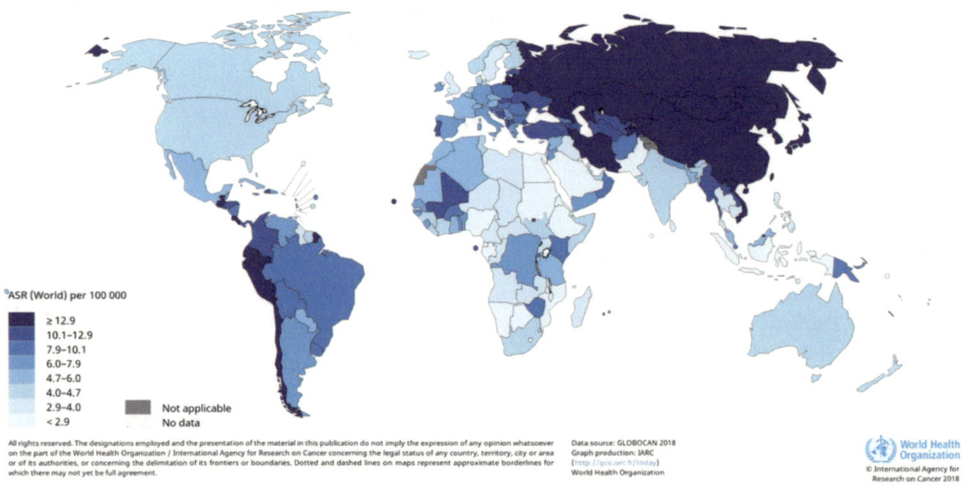

Figure 2.
Gastric cancer age-standardized rate incidence worldwide (GLOBOCAN 2018).

1.2 Classification

GC classification has been changing according to its anatomical, histological, or molecular features without a consensus regarding the best system combining prognosis and high practicality in clinical diagnosis [10–16]. For many years, the anatomical classification was used to distinguish cardia and non-cardia GC, which have distinct etiological and epidemiological characteristics [3, 6, 11]. Two classification systems that have been used for diagnosis and treatment decisions are the *Lauren classification* and *World Health Organization* (WHO) *classification*; nevertheless, the clinical impact is still limited [10, 17].

The *Lauren classification* divides gastric adenocarcinomas into diffuse, intestinal, and intermediate type, which combines cancers with uncommon histology [10, 18, 19]. There are multiple evidences indicating that the two principal subtypes may have distinct tumor development pathways [10, 18]. The intestinal type presents specific characteristics such as well/moderate differentiation of cells, loss of E-cadherin expression and is associated with H. *pylori* infection [7, 20]. The carcinogenesis model of this subtype is characterized by a progressive model characterized by chronic gastritis and gastric mucosa metaplasia [10]. The diffuse type is characterized by poorly differentiated cells with cellular atypia and numerous mitotic figures and poorly cohesive structure, and therefore it is more aggressive and with worse prognosis [11, 18, 19]. The WHO classification divides GC according to the histological features of each subtype: papillary, tubular and mucinous adenocarcinomas, poorly cohesive (including signet-ring cell carcinomas), mixed carcinomas (with two or more components), and uncommon variants [21, 22].

In 2014, *The Cancer Genome Atlas* (TCGA) consortium group proposed a classification of gastric adenocarcinomas into four distinct subtypes based on their molecular features, which may have a higher clinical impact in treatment prediction and prognosis: (1) microsatellite unstable tumors (MSI), (2) genomically stable tumors (GS), (3) tumors with chromosomal instability (CIN), and (4) tumors positive for Epstein-Barr virus (EBVaGC) [17, 22, 23] (**Figure 3**). Later in 2018, Hinoue et al. described another subtype of GC, characterized by hypermutated status with single-nucleotide variants (hypermutated-SNV, HM-SNV) [24, 25]. This system seems to have a higher clinical impact in treatment prediction and prognosis when compared with previous classification systems [26, 27]. Later, the *Asian Cancer Research Group (ACRG)* has proposed a new classification according to patterns of molecular alterations, disease progression, and prognosis: (1) high microsatellite instable (MSI-high) tumors, (2) microsatellite stable with epithelial-to-mesenchymal transition (MSS/EMT) phenotype tumors, (3) microsatellite stable with *TP53*

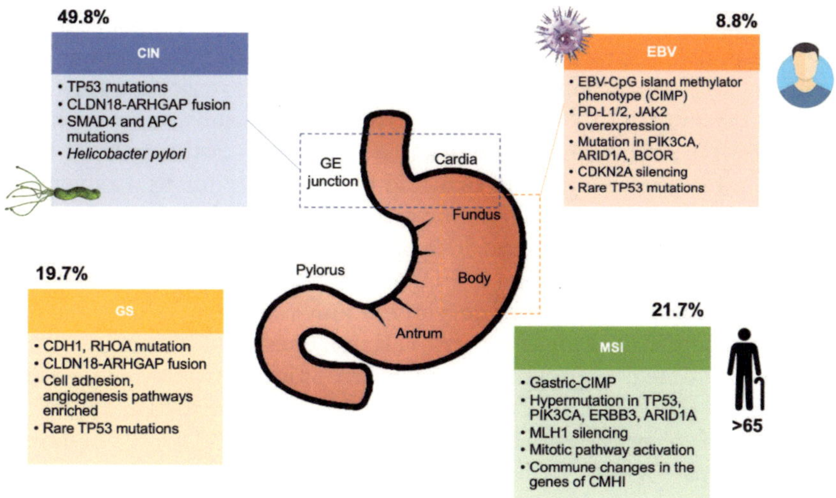

Figure 3.
Essential features of gastric cancer subtypes according to the Cancer genome atlas research network.

intact (MSS/TP53+), and (4) microsatellite stable with *TP53* loss (MSS/TP53-) [16, 28]. It is possible to obtain a partial correspondence between the TCGA and ACRG classifications, although EBV is not specifically included in the ACRG classification, EBV infection was frequently observed in the MSS/TP53+ subtype [29].

Additional subtypes of GC have been described based on the TCGA and ACRG classifications and specific analysis of different genetic features [29–31]. Nevertheless, independently of the classification system, EBV-positive GCs are considered to be of better prognosis [26, 27].

1.3 Gastric carcinogenesis and risk factors

Gastric cells' malignant transformation is a multistep process in which risk factors, genetic or epigenetic alterations can be observed [32, 33]. The carcinogenesis model for the other GC subtypes still remains a challenge for scientists due to the different histological subtypes [18, 34, 35]. The most accepted hypothesis of gastric carcinogenesis has been described for the intestinal subtype according to the *Lauren's* classification, and it is characterized by a cascade of progression from normal gastric epithelium through chronic gastritis (CG), chronic atrophic gastritis (CAG), and intestinal metaplasia (IM), ultimately leading to dysplasia and carcinoma [36, 37].

There are common risk factors for GC that can be subdivided into modifiable and non-modifiable. The non-modifiable factors include age, male gender and familiar history, and inherited syndromes, such as familial adenomatous polyposis (FAP) and Peutz-Jeghers syndrome (PJS), hereditary diffuse gastric cancer (HDGC) or Lynch syndrome [7, 38–40]. Host genetic polymorphisms have also been described to contribute to an increased risk pattern for GC development [41].

The modifiable risk factors can be divided in two major groups: dietary/lifestyle influences and infectious agents [39]. Dietary and lifestyle risk factors for GC include salt and salted preserved food, fruits and vegetables, tobacco, alcohol and body mass index (BMI), and physical activity [38, 42]. Data suggest that high salt consumption is responsible for a two-fold increase in the risk of GC development when compared to low salt intake, mainly because it induces early atrophic gastritis [39, 43]. Conversely, consumption of fresh fruits and vegetables, with vitamins C and E, carotenoids, and selenium has been associated with a decreased risk of GC in around 20-30% [7, 20]. As in other types of cancer, studies suggest that tobacco smoking is responsible for a 1.5-fold increased relative risk of developing GC [7]. Despite no explicit association, alcohol consumption is also associated with increase in risk of gastric cancer [7, 44]. A meta-analysis study has shown that high body mass index (BMI) (>25) increases the risk to develop non-cardia gastric cancer, which is 1.4-fold for overweight and two-fold in obese individuals. Conversely, regular physical activity seems to be associated with lower risk of GC [39, 45].

H. pylori infection affects around 50% of world population and has been classified by World Health organization (WHO) as a class I carcinogen being responsible for a two-fold increase in the risk of developing non-cardia gastric adenocarcinoma [7, 46, 47]. H. pylori contributes to gastric carcinogenesis by inducing chronic gastritis that over time may progress to severe atrophic gastritis, which in turn can develop to cancer [20, 44]. Other risk factors have been described as contributing to increase the risk of persistent H. pylori infection and therefore to GC development [48]. The Epstein-Barr virus [49] is another infectious agent accepted as associated with gastric carcinogenesis, however, the mechanism of action in gastric carcinogenesis is still unknown [44, 50, 51].

2. EBV-associated gastric cancer

2.1 Historical background

Epstein-Barr virus [49] is linear, double-stranded DNA virus member of herpesviridae family, with a high prevalence worldwide (>90% of adults) [52]. EBV is recognized for establishing a latent infection with frequent reactivations and has been associated with the development of multiple diseases from infectious mononucleosis to different cancers [52–55]. EBV was the first virus to be recognized as the etiological cause of a human cancer, and since 1997, it is included by the *International Agency for Research on Cancer* in the Group-I carcinogen risk factors [56–58]. Indeed, EBV has been associated with several human tumors, including Burkitt's lymphoma, Hodgkin's disease, B cell lymphomas, and also some epithelial neoplasms such as nasopharyngeal carcinoma (NPC) or more recently GC [52, 59, 60].

The association between EBV and gastric carcinoma was first described in 1990, when Burke et al. used a polymerase chain reaction (PCR) technique to detect EBV in lymphoepithelioma-like gastric carcinomas, characterized by the presence of cells morphologically similar to the undifferentiated nasopharyngeal lymphoepithelioma [61]. Later, Shibata and colleagues have demonstrated by in situ hybridization that EBV infection was present in gastric carcinoma cells resembling lymphoepithelioma but not in reactive lymphoid infiltrate or normal mucosa [62]. Additionally, 1 year later, EBV infection was also detected in cases of typical gastric adenocarcinoma [62]. Over the past 30 years, GC has been consistently associated with EBV infection [51, 61, 63, 64] and EBV-associated gastric carcinoma (EBVaGC) which is now recognized as one specific subtype of GC [51, 65, 66].

2.2 Epidemiology and clinicopathological characteristics

Several meta-analyses have been attempting to summarize the association of EBV with CG development, showing that EBVaGC may represent almost 10% of all gastric cancer worldwide [66–74]. A recent meta-analysis with data from over 20,000 cancer patients within 26 different countries has shown that EBVaGC prevalence ranges from 1.69 to 43.75%, with a pooled prevalence of 8.77% (95% CI: 7.73–9.92%) and a pooled odds ratio (OR) of 18.56 (95% CI: 15.68–21.97) for studies with matched pairs and 3.31 (95% CI: 0.95–11.54) for studies with non-matched pairs design [72].

In contrast to others EBV-associated malignancies, EBVaGC is a non-endemic endemic disease distributed throughout the world. Data analysis showed no significant variation within the different world regions (8.21% in Europe, 8.38% in Asia, 9.51% in America, and 11.9% in Africa); indeed, it is possible to observe similar data in countries from Europe, including Portugal (8,4%) [75], Netherlands (7.8%) [67], and Denmark (7.6%) [76], and also from Asiatic countries such as South Korea (7.8%) [77] and Japan (8.0%) [78]. Nevertheless, it is still possible to observe differences among countries, and the differential expositions to risk factors are proposed as the possible explanation for the variation of EBVaGC prevalence [68, 69]. Indeed, some studies suggest that EBVaGC prevalence might be inversely correlated with the background incidence of GC [70].

EBVaGC seems to be more prevalent in younger patients and in males than in females (almost two-fold more prevalent), which has been a consistent association found in several, suggesting a potential association with lifestyle or hormonal factors [68, 69, 72, 79–84] (**Figure 4**). In addition, EBVaGC seems to be more frequently found in the proximal stomach and has a moderate to poor degree of differentiation

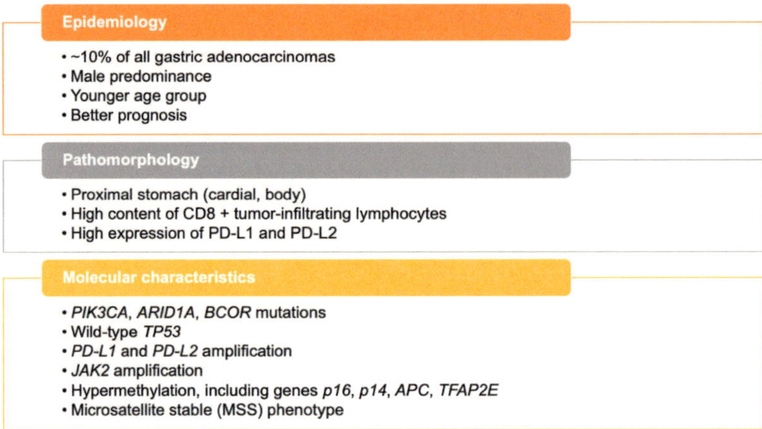

Figure 4.
Features of Epstein-Barr virus-associated gastric carcinomas.

[17, 58, 69, 72, 81, 82, 85–87]. These features are being suggested as potentially impacting the overall survival and recurrence of EBVaGC [26, 66, 80, 88], and a study with 4599 patients (pooled analysis) showed that EBVaGC has increased the overall survival and this is still a controversial topic [80]. One study evaluated the clinical significance of the different molecular subtypes of gastric cancer and concluded that EBVaGC, independently of the classification system, is of better prognosis [26, 27].

EBV association with gastric cancer was first described in lymphoepithelioma-like carcinoma also known as carcinoma with lymphoid stroma (GCLS) or medullary carcinomas [61, 88–90]. Recently, it has been described that there are three histological subtypes of EBVaGC based on the host cellular immune response status: the carcinomas with lymphoid stroma (GCLS), the carcinoma with Crohn's disease-like lymphoid reaction (CLR), and the conventional-type adenocarcinoma (CA) [91].GCLS is a rare histological subtype of gastric cancer, representing about 1-4% of all gastric cancers, of which literature shows that more than 90% of these cases are EBV-positive and characterized by poorly differentiated nests of neoplastic epithelial cells intermingled with a dense lymphoid proliferation [61, 63, 66, 69, 90, 92–94]. Nevertheless, literature has been focusing on the characterization of non-GCLs EBV-positive gastric cancers. This association remains controversial, and while several studies demonstrated a strong EBV association with a diffuse subtype [68, 95], others have reported a similar prevalence between intestinal and diffuse subtypes [67, 88, 96, 97]. Indeed, one meta-analysis has shown association with a diffuse subtype [66, 68], while two other meta-analyses did not find any association within histological subtypes [69, 74].

2.3 Diagnosis

The identification of EBVaGC has been performed by the identification of EBV transcripts in gastric tissues using in situ hybridization (ISH) by detecting EBV-encoded small RNAs (EBERs), which are highly expressed in latently EBV-infected cells [66, 91]. EBV-associated tumors are defined as monoclonal proliferations of carcinoma cells with latent EBV infection, and studies have confirmed that every cell from the cancer clone carries the clonal virus genome, suggesting that the virus was acquired before the transformation, even though it seems that it is not detected in precursor lesions [70, 86, 98].

The detection of EBV by PCR-based methods has been controversial since it frequently provides false positive results due to the presence of EBV-positive lymphocytes in the surrounding tissue, ignoring that it might be absent in the tumor epithelial cells [66]. Therefore, EBER-ISH is considered the gold-standard method, and a positive EBV-associated case should be considered only if in the presence of EBERs in tumor cells and in its absence in the normal surrounding tissue [99].

2.4 Carcinogenesis mechanism

During the past decade, several authors have been discussing the mechanism of EBV carcinogenesis in GC [68, 100]. EBV is known to enter cells in oropharyngeal lymphoid tissue by the recognition/interaction with CR2/CD21 on the surface of B-lymphocytes that interact with EBV envelope glycoprotein gp350 [101, 102]. How and when EBV gets into gastric epithelial cells remains unclear, and it has been suggested that it can be either by cell-to-cell with B-lymphocytes recruited in inflammatory processes of gastric mucosa or through direct entry into the gastric epithelia [103]. This mechanism is not well understood and further studies should be made to establish if the recruitment of EBV-infected lymphoid cells might be the explanation for the infection and subsequent transformation of gastric epithelium.

Overall, literature suggests that EBV participates on gastric carcinogenesis by both direct and indirect mechanisms: infecting epithelial cells and establishing a latent program in which a restrict profile of latent proteins/transcripts are expressed; and/or promoting a chronic inflammatory response contributing to tissue damage and cancer progression [104, 105].

Previous studies regarding the detection of EBV in premalignant lesions of gastric cancer are extremely controversial [106–109] and the majority report its presence mainly in dysplasia and atrophic gastritis adjacent to tumors [87, 105–114]. A recent cross-sectional study from the North Region of Portugal showed no evidence of EBV infection in both dysplasia and early gastric carcinomas [75]. The absence of EBER transcripts in superficial gastric neoplastic lesions may suggest that EBV infection is a late event in gastric carcinogenesis [75]. Hence, it is still important to clarify the moment of EBV infection in gastric cells and if it acts as the initiator of carcinogenesis or as a promoter after prior modifications of gastric cells.

EBVaGCs are EBV-associated epithelial malignancies and therefore the mechanism of viral carcinogenesis might be similar to the observed in NPC. Two *in vitro* studies demonstrated that nasopharyngeal cells need to have some genetic change prior to be susceptible of EBV transformation [115, 116]. In fact, preexisting genetic events, mainly cyclin D1 overexpression and p16 mutations, seem to support the establishment of stable EBV infection and transformation in NPC epithelium [115, 116]. A recent publication suggests that EBV coordinates with somatic gene mutations in order to induce the carcinogenesis process in gastric epithelial cells [117]. This mechanism suggests that high-frequency mutations, such as in PIK3CA and ARID1A, are essential for the transformation of normal gastric cells into susceptible cells, which are more likely to be infected and transformed by EBV [117]. In addition, after infection, amplification of PD-L1 and PD-L2 are thought to increase the progression and immune evasion of transformed cells [117].

Some studies have been suggesting a possible interaction between H. pylori and EBV in gastric cancer development. Minoura-Etoh et al. observed a possible antagonism effect between H. pylori and EBV, showing that reactive products from H. pylori seem to induce EBV reactivation from latently in infected gastric epithelial cells, which would avoid the EBV transformation of gastric cells in the same areas of H. pylori colonization [118]. H. pylori seems to preferentially colonize the antral region, while EBV is more frequently found in the upper third and middle of stomach, suggesting a possible antagonism of EBV and H. pylori in gastric mucosa

[119–121]. By contrast, two other studies have suggested that H. pylori may contribute for EBV-associated gastric carcinogenesis by causing gastritis that perhaps might recruit EBV-carrying lymphocytes to the stomach wall, where the virus could be induced to replicate and infect gastric epithelial cells [122, 123]. Moreover, the gastric inflammation may also promote a cytokine-rich microenvironment, supporting a clonal growth of EBV-infected epithelial cells [110].

EBV establishes a latent infection allowing it to be maintained inside cells and to use the host machinery to express their own genes, regulating the cell behavior and escaping the immune system recognition [124, 125]. EBV latency is characterized by the expression of different viral proteins such as EBV nuclear antigens (EBNAs 1, 2, 3A, 3B, 3C, and EBNA-LP), EBV-encoded small RNAs (EBERs) 1 and 2, latent membrane protein (LMP 1, 2A, and 2B), and microRNAs from BamHI-A rightward transcripts, known as BARTs [58]. Depending on the infected cell type and differentiation status, different proteins are expressed, originating different latency profiles [124–126] (**Table 1**). Literature refers that the majority of EBVaGC cases show a latency II-like pattern (44%), defined by expression of EBNA1, EBERs, BARF1, and LMP2A genes, and latency I (42.9%) restricted to EBNA1, EBERs, and BARTs expression [75, 127]. The fact that different latency states seem to be associated to different malignancies explains the different mechanism of carcinogenesis on which EBV is involved and is thought to be important for EBVaGC characterization [124–126].

The function of the different EBV latent proteins has been widely studied. Each protein seems to have a significant role for the EBV cell cycle and transformation: EBNA1, expressed in every single infected cell, acts as a transcription factor responsible for the episomal maintenance, DNA replication, and indirectly to cell transformation [56, 65, 128, 129]; EBNA2 is early expressed in recently infected B cells, playing a crucial role in these cells' immortalization through the transcription of both viral and host genes [56, 130]; the EBNA3 protein family activates the transcription of cellular and viral genes, leading to the disruption of cell cycle checkpoints on different levels [56, 130]; LMP1, the major EBV oncogene, is essential for B-lymphocytes transformations, induction of apoptotic genes, epithelial cells transformations, and invasiveness and avoids cells apoptosis by different pathways [56, 125, 129, 131]; LMP2 essentially avoids the activation of the EBV lytic cycle in B-lymphocytes and modulates epithelial cell growth [56, 125, 129]; and EBER's role is not yet well understood but is thought to contribute to B cell transformation, acting as signaling and transcription factor regulators [56, 130]. EBV also encodes around

	EBNA1	EBNA2	EBNA3	LMP1	LMP2	EBERs	Malignancies
Latency I	x					x	Burkitt's lymphoma and gastric cancer
Latency II	x			x	x	x	Nasopharyngeal carcinoma, Hodgkin lymphoma, and T-cell non-Hodgkin lymphoma
Latency II-like	x				x	x	Gastric cancer
Latency III	x	x	x	x	x	x	Post-transplant lymphoma and AIDS-associated lymphoma

Table 1.
EBV-associated diseases' latency profiles.

40 miRNAs, which are known to bind and possibly participate in the regulation of hundreds of cellular and viral transcripts, some of them being involved in cell survival [56, 130]. These miRNAs can be referred as BHRF1 and BARTs, depending on its localization on the viral genome [130]. BHRF1 role is not very understood yet but some results suggest that BHRF1 miRNAs and proteins cooperate to control cell cycle initiation and apoptosis during primary infection [130]. Regarding BARTs, they are described as contributing to EBV-induced carcinogenesis by downregulating host genes, such as tumor suppressors and pro-apoptotic genes, including several cell growth and cell cycle-related [129, 130]. Nevertheless, is still important to clarify the coordination of virus and host cell in gastric cancer carcinogenesis.

2.5 Molecular features of EBVaGC

EBVaGC has some distinctive features in terms of genome alterations [17, 66, 132] (**Figure 4**). EBVaGC has been reported to have the most extensive CpG island methylation (human and viral genomes) than in any other tumor. This is described as EBV-CIMP (CpG island methylator phenotype) and includes genes related to cell cycle regulation (p14ARF, p15, p16INK4A, and p73), DNA repair (hMLH1, MGMT and GSTP1), cell adhesion and metastases (CDH1, TIMP1, and TIMP3), apoptosis (DAPK and bcl-2), and signal transduction (APC, PTEN, and RASSF1A) [17, 133–135].

EBVaGC is characterized by mutations in the PIK3CA gene and amplification of 9p24.1 locus containing JAK2, CD274, PDCD1LG2, and ERBB2 which contribute to altered proliferation, deregulation of apoptosis, and immune suppression and evasion [17, 136]. PIK3CA gene, which encodes phosphatidyl inositol-3-kinase (PIK3), has been consistently shown to be mutated in EBVaGC [17]. This protein is an important component of PI3K/Akt/mTOR signaling pathway and regulates several cellular processes such as apoptosis escape, cell growth, and proliferation [137]. Mutations in *PIK3CA* are common in several tumors, nevertheless in EBVaGC, about 80% are non-silent mutations and the vast majority are not located in the hot-spot sites but are dispersed in the gene sequence [17, 137]. EBVaGC has also been described as having mutations in other genes such as *ARID1A* and *BCOR* [17, 66, 132]. Interesting, *TP53* mutations that occur in the majority of gastric tumors are rare in EBVaGC, nevertheless a study has shown that these tumors seem to present a lower level of *TP53* mRNA and a higher level of p53 protein when compared with EBV-negative cancers, which increases the interest in studying the p53 pathway regulation [17, 138].

EBVaGC has also been described to have higher levels of programmed death ligands 1 and 2 (PD-L1/2) enriched with CD8 + tumor-infiltrating lymphocytes (TILs) and with high expression of immunogenic pathways [25, 139, 140]. Indeed, this is considered a highly immunogenic tumor with a great potential for immunotherapy [24].

2.6 Treatment options for EBVaGC

The unique molecular features of EBVaGC have gained interest in the past years, especially for the potential impact of targeted drugs since preclinical data have shown that EBVaGC is resistant to current chemotherapy [141].

PD-L1 overexpression has been consistently considered a marker for EBVaGC and MSI-high GC cases [142, 143]. Several PD-1 targeted drugs available on the market are being studied for its use in several cancers, including GC [141]. Pembrolizumab, a PD-1 antibody, was the first to be approved by the Food and Drug Administration (FDA) for use in recurrent MSI-high GC after a good rate response in several clinical trials (NCT03257163, NCT02589496) [142, 144–146]. Pembrolizumab has also been used for the treatment of EBV-positive T cell lymphomas [147] and trials with EBVaGC are showing promising results [142]. Several clinical trials that include EBVaGC are testing other PD-1 target drugs, such as nivolumab (NCT02951091) or avelumab (NCT01772004), or by using CRISPR-Cas9-mediated PD-1 knockout EBV cytotoxic T cells (NCT03044743) [148–150]. Despite this, there are some controversial points regarding PD-L1 standardization and cutoffs, and the results from these studies point for an important role as a therapeutical target for GC, particularly in those with MSI-high or EBV. Indeed, EBV is now considered a biomarker for GC and the clear identification of EBVaGC in clinical series will contribute for the implement of better treatment strategies.

Literature shows that EBaGC has frequent *PIK3CA* mutations [17] and is thought to impact negatively the outcome of disease; nevertheless, the impact on the evolution of these cancers is still to understood [151–153]. PI3K/AKT/mTOR pathway inhibitors have been used as new therapeutical options in cancer, especially mTOR inhibitors such as everolimus, which are used in the phase III GRANITE-1 study (NCT00879333) for advanced GC with potential interest. More recently, PI3K inhibitors such as buparlisib (BKM120) have been tested for use in solid tumors [154, 155], and alpelisib has been tested for use as a potential therapeutical agent for gastric cancer [156]. There are a lot of PI3K/AKT/mTOR pathway inhibitors being used in GC clinical trials, and despite not being directed to EBVGC, a potential impact in this specific subgroup is expected.

Another important feature with potential therapeutical interest is the epigenetic changes of EBVaGC. It is known that epigenetic changes are reversible and therefore many de-methylating agents are been studied in cancer treatment. A few studies have reported on the impact of 5-azacitidine or trichostatin A in the activation of EBV lytic phase in EBVaGC cell lines, leading to the lysis of tumor cells [157–160]. Despite the potential interest, it is important to clearly understand the mechanisms of EBV lytic phase activation using de-methylating agents.

3. Conclusions

EBV has been consistently associated with GC development for almost 30 years until in 2014, when The Cancer Genome Atlas Research Network recognized EBVaGC as a specific subtype of GC. Overall, EBVaGC represents almost 10% of all gastric cancer worldwide with a prevalence variation according to geographic and risk factors exposition. EBVaGC is more prevalent in males and younger patients and is frequently found in the proximal stomach. These tumors have a moderate to poor degree of differentiation and are characterized by a high content of CD8 + tumor-infiltrating lymphocytes and high expression of PD-L1 and PD-L2 and are therefore of great potential for immunotherapy. Indeed, EBVaGC seems to has a better prognosis and increased overall survival.

EBVaGC has distinctive molecular features: (1) extensive CpG island methylation (human and viral genomes) being described as EBV-CIMP (CpG island methylator phenotype); (2) mutations in *PIK3CA*, *ARID1A*, and *BCOR* genes; (3) amplification of 9p24.1 locus containing *JAK2*, *CD274*, *PDCD1LG2*, and *ERBB2*; (4) absence of *TP53* mutations; and (5) a microsatellite stable (MSs) phenotype. The development of therapeutical approaches directed to these specific molecular features (anti-PD1, PI3K/AKT/mTOR pathway inhibitors, or demethylating drugs) is expected to impact the GC management significantly.

In sum, EBVaGC is a specific subtype of GC, presenting special clinical and pathological characteristics that could be used for the development new potential therapeutical approaches, making this an important topic for the future of gastrointestinal tumors.

Author details

Hugo Manuel Lopes de Sousa[1,2]*, Joana Patrícia Costa Ribeiro[3] and Mafalda Basílio Timóteo[2]

1 Virology Service, Portuguese Oncology Institute of Porto (IPO Porto), Rua Dr. António Bernardino de Almeida, Porto, Portugal

2 Molecular Oncology and Viral Pathology Group (CI-IPOP), Portuguese Oncology Institute of Porto (IPO Porto), Rua Dr. António Bernardino de Almeida, Porto, Portugal

3 Molecular Haematology Department, Oxford University NHS Foundation Trust, Level 4, John Radcliffe Hospital, Oxford, UK

*Address all correspondence to: hugo.sousa@ipoporto.min-saude.pt

References

[1] Ferlay J, Soerjomataram I, Ervik M, Dikshit R, Eser S, Mathers C, et al. GLOBOCAN 2012 v1.0, Cancer Incidence and Mortality Worldwide: IARC CancerBase No. 11 [Internet]. Lyon, France: International Agency for Research on Cancer; 2013. Available from: http://globocan.iarc.fr

[2] Ferlay J, Steliarova-Foucher E, Lortet-Tieulent J, Rosso S, Coebergh JW, Comber H, et al. Cancer incidence and mortality patterns in Europe: Estimates for 40 countries in 2012. European Journal of Cancer. 2013;**49**(6):1374-1403

[3] Bray F, Ferlay J, Soerjomataram I, Siegel RL, Torre LA, Jemal A. Global cancer statistics 2018: GLOBOCAN estimates of incidence and mortality worldwide for 36 cancers in 185 countries. CA: A Cancer Journal for Clinicians. 2018;**68**(6):394-424

[4] RORENO. Registo Oncológico National 2010. Porto: Instituto Português de Oncologia do Porto Francisco Gentil - EPE; 2016

[5] Fock KM. Review article: The epidemiology and prevention of gastric cancer. Alimentary Pharmacology & Therapeutics. 2014;**40**(3):250-260

[6] Balakrishnan M, George R, Sharma A, Graham DY. Changing trends in stomach cancer throughout the world. Current Gastroenterology Reports. 2017;**19**(8):36

[7] Ang TL, Fock KM. Clinical epidemiology of gastric cancer. Singapore Medical Journal. 2014;**55**(12):621-628

[8] Torre LA, Bray F, Siegel RL, Ferlay J, Lortet-Tieulent J, Jemal A. Global cancer statistics, 2012. CA: A Cancer Journal for Clinicians. 2015;**65**(2):87-108

[9] Collaborators GBDSC. The global, regional, and national burden of stomach cancer in 195 countries, 1990-2017: A systematic analysis for the global burden of disease study 2017. The Lancet Gastroenterology & Hepatology. 2020;**5**(1):42-54

[10] Berlth F, Bollschweiler E, Drebber U, Hoelscher AH, Moenig S. Pathohistological classification systems in gastric cancer: Diagnostic relevance and prognostic value. World Journal of Gastroenterology. 2014;**20**(19):5679-5684

[11] Van Cutsem E, Sagaert X, Topal B, Haustermans K, Prenen H. Gastric cancer. Lancet. 2016;**388**(10060):2654-2664

[12] Grundmann E, Schlake W. Histological classification of gastric cancer from initial to advanced stages. Pathology, Research and Practice. 1982;**173**(3):260-274

[13] Carneiro F, Seixas M, Sobrinho-Simoes M. New elements for an updated classification of the carcinomas of the stomach. Pathology, Research and Practice. 1995;**191**(6):571-584

[14] Goseki N, Takizawa T, Koike M. Differences in the mode of the extension of gastric cancer classified by histological type: New histological classification of gastric carcinoma. Gut. 1992;**33**(5):606-612

[15] Ming SC. Gastric carcinoma. A pathobiological classification. Cancer. 1977;**39**(6):2475-2485

[16] Cislo M, Filip AA, Arnold Offerhaus GJ, Cisel B, Rawicz-Pruszynski K, Skierucha M, et al. Distinct molecular subtypes of gastric cancer: From Lauren to molecular pathology. Oncotarget. 2018;**9**(27):19427-19442

[17] Cancer Genome Atlas Research N. Comprehensive

molecular characterization of gastric adenocarcinoma. Nature. 2014;**513**(7517):202-209

[18] Lauren P. The two histological main types of gastric carcinoma: Diffuse and so-called intestinal-type carcinoma. An attempt at a histo-clinical classification. Acta Pathologica et Microbiologica Scandinavica. 1965;**64**:31-49

[19] Marques-Lespier JM, Gonzalez-Pons M, Cruz-Correa M. Current perspectives on gastric cancer. Gastroenterology Clinics of North America. 2016;**45**(3):413-428

[20] Kelley JR, Duggan JM. Gastric cancer epidemiology and risk factors. Journal of Clinical Epidemiology. 2003;**56**(1):1-9

[21] Bosman FT, Carneiro F, Hruban RH, Theise ND. WHO classification of tumours of the digestive system. In: World Health Organization Classification of Tumours. 4th edition. The International Agency for Research on Cancer; 2010

[22] Rocken C. Molecular classification of gastric cancer. Expert Review of Molecular Diagnostics. 2017;**17**(3):293-301. DOI: 10.1080/14737159.2017.1286985

[23] Sunakawa Y, Lenz HJ. Molecular classification of gastric adenocarcinoma: Translating new insights from the cancer genome atlas research network. Current Treatment Options in Oncology. 2015;**16**(4):17

[24] Ignatova E, Seriak D, Fedyanin M, Tryakin A, Pokataev I, Menshikova S, et al. Epstein-Barr virus-associated gastric cancer: Disease that requires special approach [published online ahead of print]. Gastric Cancer. 8 June 2020. DOI:10.1007/s10120-020-01095-z

[25] Liu Y, Sethi NS, Hinoue T, Schneider BG, Cherniack AD, Sanchez-Vega F, et al. Comparative molecular analysis of gastrointestinal adenocarcinomas. Cancer Cell. 2018;**33**(4):721-735 e8

[26] Sohn BH, Hwang JE, Jang HJ, Lee HS, Oh SC, Shim JJ, et al. Clinical significance of four molecular subtypes of gastric cancer identified by the Cancer Genome Atlas project [published online ahead of print]. Clinical Cancer Research. 26 July 2017. DOI: 10.1158/1078-0432.CCR-16-2211

[27] Kim Y, Cho MY, Kim J, Kim SN, Oh SC, Lee KA. Profiling cancer-associated genetic alterations and molecular classification of cancer in Korean gastric cancer patients. Oncotarget. 2017;**8**(41):69888-69905

[28] Cristescu R, Lee J, Nebozhyn M, Kim KM, Ting JC, Wong SS, et al. Molecular analysis of gastric cancer identifies subtypes associated with distinct clinical outcomes. Nature Medicine. 2015;**21**(5):449-456

[29] Wang Q, Liu G, Hu C. Molecular classification of gastric adenocarcinoma. Gastroenterology Research. 2019;**12**(6):275-282

[30] Setia N, Agoston AT, Han HS, Mullen JT, Duda DG, Clark JW, et al. A protein and mRNA expression-based classification of gastric cancer. Modern Pathology. 2016;**29**(7):772-784

[31] Ahn S, Lee SJ, Kim Y, Kim A, Shin N, Choi KU, et al. High-throughput protein and mRNA expression-based classification of gastric cancers can identify clinically distinct subtypes, concordant with recent molecular classifications. The American Journal of Surgical Pathology. 2017;**41**(1):106-115

[32] Hu XT, He C. Recent progress in the study of methylated tumor suppressor genes in gastric cancer. Chinese Journal of Cancer. 2013;**32**(1):31-41

[33] Qiu T, Zhou X, Wang J, Du Y, Xu J, Huang Z, et al. MiR-145, miR-133a

and miR-133b inhibit proliferation, migration, invasion and cell cycle progression via targeting transcription factor Sp1 in gastric cancer. FEBS Letters. 2014;**588**(7):1168-1177

[34] Dinis-Ribeiro M, Areia M, de Vries AC, Marcos-Pinto R, Monteiro-Soares M, O'Connor A, et al. Management of precancerous conditions and lesions in the stomach (MAPS): Guideline from the European Society of Gastrointestinal Endoscopy (ESGE), European Helicobacter Study Group (EHSG), European Society of Pathology (ESP), and the Sociedade Portuguesa de Endoscopia Digestiva (SPED). Endoscopy. 2012;**44**(1):74-94

[35] Tan VP, Wong BC. *Helicobacter pylori* and gastritis: Untangling a complex relationship 27 years on. Journal of Gastroenterology and Hepatology. 2011;**26**(Suppl 1):42-45

[36] Dixon MF, Genta RM, Yardley JH, Correa P. Classification and grading of gastritis. The updated Sydney System. International Workshop on the Histopathology of Gastritis, Houston 1994. The American Journal of Surgical Pathology. 1996;**20**(10):1161-1181

[37] Dixon MF. Gastrointestinal epithelial neoplasia: Vienna revisited. Gut. 2002;**51**(1):130-131

[38] Karimi P, Islami F, Anandasabapathy S, Freedman ND, Kamangar F. Gastric cancer: Descriptive epidemiology, risk factors, screening, and prevention. Cancer Epidemiology, Biomarkers & Prevention: A Publication of the American Association for Cancer Research, Cosponsored by the American Society of Preventive Oncology. 2014;**23**(5):700-713

[39] de Martel C, Forman D, Plummer M. Gastric cancer: Epidemiology and risk factors. Gastroenterology Clinics of North America. 2013;**42**(2):219-240

[40] Goodenberger M, Lindor NM. Lynch syndrome and MYH-associated polyposis: Review and testing strategy. Journal of Clinical Gastroenterology. 2011;**45**(6):488-500

[41] Ramos M, Ribeiro Junior U, Viscondi JKY, Zilberstein B, Cecconello I, Eluf-Neto J. Risk factors associated with the development of gastric cancer—Case-control study. Revista da Associação Médica Brasileira (1992). 2018;**64**(7):611-619

[42] Fang X, Wei J, He X, An P, Wang H, Jiang L, et al. Landscape of dietary factors associated with risk of gastric cancer: A systematic review and dose-response meta-analysis of prospective cohort studies. European Journal of Cancer. 2015;**51**(18):2820-2832

[43] Ge S, Feng X, Shen L, Wei Z, Zhu Q, Sun J. Association between habitual dietary salt intake and risk of gastric cancer: A systematic review of observational studies. Gastroenterology Research and Practice. 2012;**2012**:808120

[44] Guggenheim DE, Shah MA. Gastric cancer epidemiology and risk factors. Journal of Surgical Oncology. 2013;**107**(3):230-236

[45] Yang P, Zhou Y, Chen B, Wan HW, Jia GQ, Bai HL, et al. Overweight, obesity and gastric cancer risk: Results from a meta-analysis of cohort studies. European Journal of Cancer. 2009;**45**(16):2867-2873

[46] Schistosomes, liver flukes and Helicobacter pylori. IARC Working Group on the Evaluation of Carcinogenic Risks to Humans. Lyon, 7-14 June 1994. IARC Monogr Eval Carcinog Risks Hum. 1994;**61**:1-241

[47] Plummer M, Franceschi S, Vignat J, Forman D, de Martel C. Global burden

of gastric cancer attributable to *Helicobacter pylori*. International Journal of Cancer. 2015;**136**(2):487-490

[48] Gaddy JA, Radin JN, Loh JT, Zhang F, Washington MK, Peek RM Jr, et al. High dietary salt intake exacerbates *Helicobacter pylori*-induced gastric carcinogenesis. Infection and Immunity. 2013;**81**(6):2258-2267

[49] Mohammad AH, Assadian S, Couture F, Lefebvre KJ, El-Assaad W, Barres V, et al. V-ATPase-associated prorenin receptor is upregulated in prostate cancer after PTEN loss. Oncotarget. 2019;**10**(48):4923-4936

[50] Piazuelo MB, Epplein M, Correa P. Gastric cancer: An infectious disease. Infectious Disease Clinics of North America. 2010;**24**(4):853-869, vii

[51] Takada K. Epstein-Barr virus and gastric carcinoma. Molecular Pathology: MP. 2000;**53**(5):255-261

[52] Young LS, Rickinson AB. Epstein-Barr virus: 40 years on. Nature Reviews. Cancer. 2004;**4**(10):757-768

[53] Javier RT, Butel JS. The history of tumor virology. Cancer Research. 2008;**68**(19):7693-7706

[54] Kutok JL, Wang F. Spectrum of Epstein-Barr virus-associated diseases. Annual Review of Pathology. 2006;**1**:375-404

[55] Murray PG, Young LS. Epstein-Barr virus infection: Basis of malignancy and potential for therapy. Expert Reviews in Molecular Medicine. 2001;**3**(28):1-20

[56] IWGotEoCRt H. Biological agents. Volume 100 B. A review of human carcinogens. IARC Monographs on the Evaluation of Carcinogenic Risks to Humans. 2012;**100**(Pt B): 1-441

[57] Proceedings of the IARC Working Group on the Evaluation of Carcinogenic Risks to Humans. Epstein-Barr virus and Kaposi's sarcoma Herpesvirus/human Herpesvirus 8. Lyon, France, 17-24 June 1997. IARC Monographs on the Evaluation of Carcinogenic Risks to Humans. 1997;**70**:1-492

[58] Naseem M, Barzi A, Brezden-Masley C, Puccini A, Berger MD, Tokunaga R, et al. Outlooks on Epstein-Barr virus associated gastric cancer. Cancer Treatment Reviews. 2018;**66**:15-22

[59] Tsao SW, Tsang CM, To KF, Lo KW. The role of Epstein-Barr virus in epithelial malignancies. The Journal of Pathology. 2015;**235**(2):323-333

[60] Young LS, Murray PG. Epstein-Barr virus and oncogenesis: From latent genes to tumours. Oncogene. 2003;**22**(33):5108-5121

[61] Burke AP, Yen TS, Shekitka KM, Sobin LH. Lymphoepithelial carcinoma of the stomach with Epstein-Barr virus demonstrated by polymerase chain reaction. Modern Pathology: An Official Journal of the United States and Canadian Academy of Pathology, Inc. 1990;**3**(3):377-380

[62] Shibata D, Weiss LM. Epstein-Barr virus-associated gastric adenocarcinoma. The American Journal of Pathology. 1992;**140**(4):769-774

[63] Shibata D, Tokunaga M, Uemura Y, Sato E, Tanaka S, Weiss LM. Association of Epstein-Barr virus with undifferentiated gastric carcinomas with intense lymphoid infiltration. Lymphoepithelioma-like carcinoma. The American Journal of Pathology. 1991;**139**(3):469-474

[64] Tokunaga M, Land CE, Uemura Y, Tokudome T, Tanaka S, Sato E. Epstein-Barr virus in gastric carcinoma.

The American Journal of Pathology. 1993;**143**(5):1250-1254

[65] Iizasa H, Nanbo A, Nishikawa J, Jinushi M, Yoshiyama H. Epstein-Barr virus (EBV)-associated gastric carcinoma. Viruses. 2012;**4**(12):3420-3439

[66] Lee JH, Kim SH, Han SH, An JS, Lee ES, Kim YS. Clinicopathological and molecular characteristics of Epstein-Barr virus-associated gastric carcinoma: A meta-analysis. Journal of Gastroenterology and Hepatology. 2009;**24**(3):354-365

[67] van Beek J, zur Hausen A, Klein Kranenbarg E, van de Velde CJ, Middeldorp JM, van den Brule AJ, et al. EBV-positive gastric adenocarcinomas: A distinct clinicopathologic entity with a low frequency of lymph node involvement. Journal of Clinical Oncology. 2004;**22**(4):664-670

[68] Camargo MC, Murphy G, Koriyama C, Pfeiffer RM, Kim WH, Herrera-Goepfert R, et al. Determinants of Epstein-Barr virus-positive gastric cancer: An international pooled analysis. British Journal of Cancer. 2011;**105**(1):38-43

[69] Murphy G, Pfeiffer R, Camargo MC, Rabkin CS. Meta-analysis shows that prevalence of Epstein-Barr virus-positive gastric cancer differs based on sex and anatomic location. Gastroenterology. 2009;**137**(3):824-833

[70] Sousa H, Pinto-Correia AL, Medeiros R, Dinis-Ribeiro M. Epstein-Barr virus is associated with gastric carcinoma: The question is what is the significance? World Journal of Gastroenterology. 2008;**14**(27):4347-4351

[71] Oh JK, Weiderpass E. Infection and cancer: Global distribution and burden of diseases. Annals of Global Health. 2014;**80**(5):384-392

[72] Tavakoli A, Monavari SH, Solaymani Mohammadi F, Kiani SJ, Armat S, Farahmand M. Association between Epstein-Barr virus infection and gastric cancer: A systematic review and meta-analysis. BMC Cancer. 2020;**20**(1):493

[73] Bae JM, Kim EH. Epstein-Barr virus and gastric cancer risk: A meta-analysis with meta-regression of case-control studies. Journal of Preventive Medicine and Public Health. 2016;**49**(2):97-107

[74] Li S, Du H, Wang Z, Zhou L, Zhao X, Zeng Y. Meta-analysis of the relationship between Epstein-Barr virus infection and clinicopathological features of patients with gastric carcinoma. Science China. Life Sciences. 2010;**53**(4):524-530

[75] Ribeiro J, Oliveira A, Malta M, Oliveira C, Silva F, Galaghar A, et al. Clinical and pathological characterization of Epstein-Barr virus-associated gastric carcinomas in Portugal. World Journal of Gastroenterology. 2017;**23**(40):7292-7302

[76] Boysen T, Friborg J, Stribolt K, Hamilton-Dutoit S, Goertz S, Wohlfahrt J, et al. Epstein-Barr virus-associated gastric carcinoma among patients with pernicious anemia. International Journal of Cancer. 2011;**129**(11):2756-2760

[77] Kim RH, Chang MS, Kim HJ, Song KS, Kim YS, Choi BY, et al. Medical history and lifestyle factors contributing to Epstein-Barr virus-associated gastric carcinoma and conventional gastric carcinoma in Korea. Anticancer Research. 2010;**30**(6):2469-2475

[78] Sukawa Y, Yamamoto H, Nosho K, Kunimoto H, Suzuki H, Adachi Y, et al. Alterations in the human epidermal growth factor receptor 2-phosphatidylinositol 3-kinase-v-Akt pathway in gastric cancer.

World Journal of Gastroenterology. 2012;**18**(45):6577-6586

[79] Truong CD, Feng W, Li W, Khoury T, Li Q, Alrawi S, et al. Characteristics of Epstein-Barr virus-associated gastric cancer: A study of 235 cases at a comprehensive cancer center in U.S.A. Journal of Experimental & Clinical Cancer Research: CR. 2009;**28**:14

[80] Camargo MC, Kim WH, Chiaravalli AM, Kim KM, Corvalan AH, Matsuo K, et al. Improved survival of gastric cancer with tumour Epstein-Barr virus positivity: An international pooled analysis. Gut. 2014;**63**(2):236-243

[81] Rymbai ML, Ramalingam VV, Samarasan I, Chandran BS, Mathew G, Jerobin J, et al. Frequency of Epstein--Barr virus infection as detected by messenger RNA for EBNA 1 in histologically proven gastric adenocarcinoma in patients presenting to a tertiary care center in South India. Indian Journal of Medical Microbiology. 2015;**33**(3):369-373

[82] Morales-Sanchez A, Fuentes-Panana EM. Epstein-Barr virus-associated gastric cancer and potential mechanisms of oncogenesis. Current Cancer Drug Targets. 2017;**17**(6):534-554

[83] Jacome AA, Lima EM, Kazzi AI, Chaves GF, Mendonca DC, Maciel MM, et al. Epstein-Barr virus-positive gastric cancer: A distinct molecular subtype of the disease? Revista da Sociedade Brasileira de Medicina Tropical. 2016;**49**(2):150-157

[84] Qiu K, Tomita Y, Hashimoto M, Ohsawa M, Kawano K, Wu DM, et al. Epstein-Barr virus in gastric carcinoma in Suzhou, China and Osaka, Japan: Association with clinico-pathologic factors and HLA-subtype. International Journal of Cancer. 1997;**71**(2):155-158

[85] Park JH, Kim EK, Kim YH, Kim JH, Bae YS, Lee YC, et al. Epstein-Barr virus positivity, not mismatch repair-deficiency, is a favorable risk factor for lymph node metastasis in submucosa-invasive early gastric cancer. Gastric Cancer. 2016;**19**(4):1041-1051

[86] Rickinson AB. Co-infections, inflammation and oncogenesis: Future directions for EBV research. Seminars in Cancer Biology. 2014;**26**:99-115

[87] Murai K, Kakushima N, Sugino T, Yoshida M, Kawata N, Tanaka M, et al. Epstein-Barr virus positivity among surgically resected intramucosal gastric cancer. Digestive Endoscopy: Official Journal of the Japan Gastroenterological Endoscopy Society. 2018;**30**(5):667-671

[88] Corvalan A, Koriyama C, Akiba S, Eizuru Y, Backhouse C, Palma M, et al. Epstein-Barr virus in gastric carcinoma is associated with location in the cardia and with a diffuse histology: A study in one area of Chile. International Journal of Cancer. 2001;**94**(4):527-530

[89] Huh CW, Jung DH, Kim H, Kim H, Youn YH, Park H, et al. Clinicopathologic features of gastric carcinoma with lymphoid stroma in early gastric cancer. Journal of Surgical Oncology. 2016;**114**(6):769-772

[90] Lim H, Park YS, Lee JH, Son DH, Ahn JY, Choi KS, et al. Features of gastric carcinoma with lymphoid stroma associated with Epstein-Barr virus. Clinical Gastroenterology and Hepatology: The Official Clinical Practice Journal of the American Gastroenterological Association. 2015;**13**(10):1738-1744 e2

[91] Shinozaki-Ushiku A, Kunita A, Fukayama M. Update on Epstein-Barr virus and gastric cancer (review). International Journal of Oncology. 2015;**46**(4):1421-1434

[92] Watanabe H, Enjoji M, Imai T. Gastric carcinoma with lymphoid stroma. Its morphologic characteristics

and prognostic correlations. Cancer. 1976;**38**(1):232-243

[93] Lim H, Lee IS, Lee JH, Park YS, Kang HJ, Na HK, et al. Clinical application of early gastric carcinoma with lymphoid stroma based on lymph node metastasis status. Gastric Cancer. 2017;**20**(5):793-801. DOI: 10.1007/s10120-017-0703-z

[94] Wang ZH, Zhao JJ, Yuan Z. Lymphoepithelioma-like gastric carcinoma: A case report and review of the literature. World Journal of Gastroenterology. 2016;**22**(10):3056-3061

[95] Carrasco-Avino G, Riquelme I, Padilla O, Villaseca M, Aguayo FR, Corvalan AH. The conundrum of the Epstein-Barr virus-associated gastric carcinoma in the Americas. Oncotarget. 2017;**8**(43):75687-75698

[96] Chang MS, Lee HS, Kim CW, Kim YI, Kim WH. Clinicopathologic characteristics of Epstein-Barr virus-incorporated gastric cancers in Korea. Pathology, Research and Practice. 2001;**197**(6):395-400

[97] Yoshiwara E, Koriyama C, Akiba S, Itoh T, Minakami Y, Chirinos JL, et al. Epstein-Barr virus-associated gastric carcinoma in Lima, Peru. Journal of Experimental & Clinical Cancer Research: CR. 2005;**24**(1):49-54

[98] Ragazzi M, Ciarrocchi A, Sancisi V, Gandolfi G, Bisagni A, Piana S. Update on anaplastic thyroid carcinoma: Morphological, molecular, and genetic features of the most aggressive thyroid cancer. International Journal of Endocrinology. 2014;**2014**:790834

[99] Nishikawa J, Iizasa H, Yoshiyama H, Shimokuri K, Kobayashi Y, Sasaki S, et al. Clinical importance of Epstein(-) Barr virus-associated gastric cancer. Cancers (Basel). 29 May 2018;**10**(6):167. DOI: 10.3390/cancers10060167

[100] Young LS, Yap LF, Murray PG. Epstein-Barr virus: More than 50 years old and still providing surprises. Nature Reviews Cancer. 2016;**16**(12):789-802

[101] Nemerow GR, Houghten RA, Moore MD, Cooper NR. Identification of an epitope in the major envelope protein of Epstein-Barr virus that mediates viral binding to the B lymphocyte EBV receptor (CR2). Cell. 1989;**56**(3):369-377

[102] Nemerow GR, Mold C, Schwend VK, Tollefson V, Cooper NR. Identification of gp350 as the viral glycoprotein mediating attachment of Epstein-Barr virus (EBV) to the EBV/C3d receptor of B cells: Sequence homology of gp350 and C3 complement fragment C3d. Journal of Virology. 1987;**61**(5):1416-1420

[103] Yue W, Zhu M, Zuo L, Xin S, Zhang J, Liu L, et al. Early pattern of Epstein-Barr virus infection in gastric epithelial cells by "cell-in-cell". Virologica Sinica. 2019;**34**(3):253-261

[104] Ohnishi N, Yuasa H, Tanaka S, Sawa H, Miura M, Matsui A, et al. Transgenic expression of *Helicobacter pylori* CagA induces gastrointestinal and hematopoietic neoplasms in mouse. Proceedings of the National Academy of Sciences of the United States of America. 2008;**105**(3):1003-1008

[105] Martinez-Lopez JL, Torres J, Camorlinga-Ponce M, Mantilla A, Leal YA, Fuentes-Panana EM. Evidence of Epstein-Barr virus association with gastric cancer and non-atrophic gastritis. Viruses. 2014;**6**(1):301-318

[106] Gulley ML, Pulitzer DR, Eagan PA, Schneider BG. Epstein-Barr virus infection is an early event in gastric carcinogenesis and is independent of bcl-2 expression and p53 accumulation. Human Pathology. 1996;**27**(1):20-27

[107] Hungermann D, Muller S, Spieker T, Lisner R, Niedobitek G,

Herbst H. Low prevalence of latently Epstein-Barr virus-infected cells in chronic gastritis. Microscopy Research and Technique. 2001;**53**(6):409-413

[108] Nogueira C, Mota M, Gradiz R, Cipriano MA, Caramelo F, Cruz H, et al. Prevalence and characteristics of Epstein–Barr virus-associated gastric carcinomas in Portugal. Infectious Agents and Cancer. 2017;**12**(41):1-8

[109] Zur Hausen A, van Rees BP, van Beek J, Craanen ME, Bloemena E, Offerhaus GJ, et al. Epstein-Barr virus in gastric carcinomas and gastric stump carcinomas: A late event in gastric carcinogenesis. Journal of Clinical Pathology. 2004;**57**(5):487-491

[110] de Souza CR, de Oliveira KS, Ferraz JJ, Leal MF, Calcagno DQ, Seabra AD, et al. Occurrence of *Helicobacter pylori* and Epstein-Barr virus infection in endoscopic and gastric cancer patients from northern Brazil. BMC Gastroenterology. 2014;**14**:179

[111] Yanai H, Takada K, Shimizu N, Mizugaki Y, Tada M, Okita K. Epstein-Barr virus infection in non-carcinomatous gastric epithelium. The Journal of Pathology. 1997;**183**(3):293-298

[112] Luqmani YA, Linjawi SO, Shousha S. Detection of Epstein-Barr virus in gastrectomy specimens. Oncology Reports. 2001;**8**(5):995-999

[113] Hirano A, Yanai H, Shimizu N, Okamoto T, Matsubara Y, Yamamoto K, et al. Evaluation of epstein-barr virus DNA load in gastric mucosa with chronic atrophic gastritis using a real-time quantitative PCR assay. International Journal of Gastrointestinal Cancer. 2003;**34**(2-3):87-94

[114] Chen ZM, Shah R, Zuckerman GR, Wang HL. Epstein-Barr virus gastritis: An underrecognized form of severe gastritis simulating gastric lymphoma. The American Journal of Surgical Pathology. 2007;**31**(9):1446-1451

[115] Tsang CM, Yip YL, Lo KW, Deng W, To KF, Hau PM, et al. Cyclin D1 overexpression supports stable EBV infection in nasopharyngeal epithelial cells. Proceedings of the National Academy of Sciences of the United States of America. 2012;**109**(50):E3473-E3482

[116] Tsang CM, Zhang G, Seto E, Takada K, Deng W, Yip YL, et al. Epstein-Barr virus infection in immortalized nasopharyngeal epithelial cells: Regulation of infection and phenotypic characterization. International Journal of Cancer. 2010;**127**(7):1570-1583

[117] Abe H, Kaneda A, Fukayama M. Epstein-Barr virus-associated gastric carcinoma: Use of host cell machineries and somatic gene mutations. Pathobiology: Journal of Immunopathology, Molecular and Cellular Biology. 2015;**82**(5):212-223

[118] Minoura-Etoh J, Gotoh K, Sato R, Ogata M, Kaku N, Fujioka T, et al. *Helicobacter pylori*-associated oxidant monochloramine induces reactivation of Epstein-Barr virus (EBV) in gastric epithelial cells latently infected with EBV. Journal of Medical Microbiology 2006;55(Pt 7):905-911.

[119] Herrera-Goepfert R, Akiba S, Koriyama C, Ding S, Reyes E, Itoh T, et al. Epstein-Barr virus-associated gastric carcinoma: Evidence of age-dependence among a Mexican population. World Journal of Gastroenterology. 2005;**11**(39):6096-6103

[120] Galetsky SA, Tsvetnov VV, Land CE, Afanasieva TA, Petrovichev NN, Gurtsevitch VE, et al. Epstein-Barr-virus-associated gastric cancer in Russia. International Journal of Cancer. 1997;**73**(6):786-789

[121] Nogueira Tde B, Artigiani RN, Herani BF, Waisberg J. *H. pylori* infection, endoscopic, histological aspects and cell proliferation in the gastric mucosa of patients submitted to roux-en-Y gastric bypass with contention ring: A cross sectional endoscopic and immunohistochemical study. Arquivos de Gastroenterologia. 2016;**53**(1):55-60

[122] Camargo MC, Kim KM, Matsuo K, Torres J, Liao LM, Morgan DR, et al. Anti-*Helicobacter pylori* antibody profiles in Epstein-Barr virus (EBV)-positive and EBV-negative gastric cancer. Helicobacter. 2016;**21**(2):153-157

[123] Matsusaka K, Funata S, Fukayama M, Kaneda A. DNA methylation in gastric cancer, related to *Helicobacter pylori* and Epstein-Barr virus. World Journal of Gastroenterology. 2014;**20**(14):3916-3926

[124] White MK, Pagano JS, Khalili K. Viruses and human cancers: A long road of discovery of molecular paradigms. Clinical Microbiology Reviews. 2014;**27**(3):463-481

[125] Raab-Traub N. EBV-induced oncogenesis. In: Arvin A, Campadelli-Fiume G, Mocarski E, Moore PS, Roizman B, Whitley R, et al., editors. Human Herpesviruses: Biology, Therapy, and Immunoprophylaxis. Cambridge: Cambridge University Press; 2007. Chapter 55. Available from: https://www.ncbi.nlm.nih.gov/books/NBK47429/

[126] Tang W, Morgan DR, Meyers MO, Dominguez RL, Martinez E, Kakudo K, et al. Epstein-Barr virus infected gastric adenocarcinoma expresses latent and lytic viral transcripts and has a distinct human gene expression profile. Infectious Agents and Cancer. 2012;**7**(1):21

[127] Ribeiro J, Oliveira C, Malta M, Sousa H. Epstein-Barr virus gene expression and latency pattern in gastric carcinomas: A systematic review. Future Oncology. 2017;**13**(6):567-579

[128] Amon W, Farrell PJ. Reactivation of Epstein-Barr virus from latency. Reviews in Medical Virology. 2005;**15**(3):149-156

[129] Raab-Traub N. Novel mechanisms of EBV-induced oncogenesis. Current Opinion in Virology. 2012;**2**(4):453-458

[130] Fitzsimmons L, Kelly GL. EBV and apoptosis: The viral master regulator of cell fate? Viruses. 13 Nov 2017;**9**(11):339. DOI: 10.3390/v9110339

[131] Curran JA, Laverty FS, Campbell D, Macdiarmid J, Wilson JB. Epstein-Barr virus encoded latent membrane protein-1 induces epithelial cell proliferation and sensitizes transgenic mice to chemical carcinogenesis. Cancer Research. 2001;**61**(18):6730-6738

[132] Yau TO, Tang CM, Yu J. Epigenetic dysregulation in Epstein-Barr virus-associated gastric carcinoma: Disease and treatments. World Journal of Gastroenterology. 2014;**20**(21):6448-6456

[133] Zouridis H, Deng N, Ivanova T, Zhu Y, Wong B, Huang D, et al. Methylation subtypes and large-scale epigenetic alterations in gastric cancer. Science Translational Medicine. 2012;**4**(156):156ra40

[134] Gulley ML. Genomic assays for Epstein-Barr virus-positive gastric adenocarcinoma. Experimental & Molecular Medicine. 2015;**47**:e134

[135] Chapel F, Fabiani B, Davi F, Raphael M, Tepper M, Champault G, et al. Epstein-Barr virus and gastric carcinoma in Western patients: Comparison of pathological parameters and p53 expression in EBV-positive and negative tumours. Histopathology. 2000;**36**(3):252-261

[136] Sun Q, Brewer N, Dunham K, Chen L, Bao L, Burton R, et al. Interferon-gamma expressing EBV LMP2A-specific T cells for cellular immunotherapy. Cellular Immunology. 2007;**246**(2):81-91

[137] Samuels Y, Waldman T. Oncogenic mutations of PIK3CA in human cancers. Current Topics in Microbiology and Immunology. 2010;**347**:21-41

[138] Ribeiro J, Malta M, Galaghar A, Silva F, Afonso LP, Medeiros R, et al. P53 deregulation in Epstein-Barr virus-associated gastric cancer. Cancer Letters. 2017;**404**:37-43

[139] Derks S, Liao X, Chiaravalli AM, Xu X, Camargo MC, Solcia E, et al. Abundant PD-L1 expression in Epstein-Barr virus-infected gastric cancers. Oncotarget. 2016;7(22):32925-32932

[140] Koh J, Ock CY, Kim JW, Nam SK, Kwak Y, Yun S, et al. Clinicopathologic implications of immune classification by PD-L1 expression and CD8-positive tumor-infiltrating lymphocytes in stage II and III gastric cancer patients. Oncotarget. 2017;**8**(16):26356-26367

[141] Kang BW, Baek DW, Kang H, Baek JH, Kim JG. Novel therapeutic approaches for Epstein-Barr virus associated gastric cancer. Anticancer Research. 2019;**39**(8):4003-4010

[142] Kim ST, Cristescu R, Bass AJ, Kim KM, Odegaard JI, Kim K, et al. Comprehensive molecular characterization of clinical responses to PD-1 inhibition in metastatic gastric cancer. Nature Medicine. 2018;**24**(9):1449-1458

[143] Gu L, Chen M, Guo D, Zhu H, Zhang W, Pan J, et al. PD-L1 and gastric cancer prognosis: A systematic review and meta-analysis. PLoS One. 2017;**12**(8):e0182692

[144] Fuchs CS, Doi T, Jang RW, Muro K, Satoh T, Machado M, et al. Safety and efficacy of pembrolizumab monotherapy in patients with previously treated advanced gastric and gastroesophageal junction cancer: Phase 2 clinical KEYNOTE-059 trial. JAMA Oncology. 2018;**4**(5):e180013

[145] Muro K, Chung HC, Shankaran V, Geva R, Catenacci D, Gupta S, et al. Pembrolizumab for patients with PD-L1-positive advanced gastric cancer (KEYNOTE-012): A multicentre, open-label, phase 1b trial. The Lancet Oncology. 2016;**17**(6):717-726

[146] Joshi SS, Maron SB, Catenacci DV. Pembrolizumab for treatment of advanced gastric and gastroesophageal junction adenocarcinoma. Future Oncology. 2018;**14**(5):417-430

[147] Kwong YL, Chan TSY, Tan D, Kim SJ, Poon LM, Mow B, et al. PD1 blockade with pembrolizumab is highly effective in relapsed or refractory NK/T-cell lymphoma failing l-asparaginase. Blood. 2017;**129**(17):2437-2442

[148] Moehler M, Ryu MH, Dvorkin M, Lee KW, Coskun HS, Wong R, et al. Maintenance avelumab versus continuation of first-line chemotherapy in gastric cancer: JAVELIN Gastric 100 study design. Future Oncology. 2019;**15**(6):567-577

[149] Bang YJ, Ruiz EY, Van Cutsem E, Lee KW, Wyrwicz L, Schenker M, et al. Phase III, randomised trial of avelumab versus physician's choice of chemotherapy as third-line treatment of patients with advanced gastric or gastro-oesophageal junction cancer: Primary analysis of JAVELIN Gastric 300. Annals of Oncology. 2018;**29**(10):2052-2060

[150] Panda A, Mehnert JM, Hirshfield KM, Riedlinger G, Damare S, Saunders T, et al. Immune activation and benefit from avelumab in EBV-positive gastric cancer. Journal of the National Cancer Institute. 2018;**110**(3):316-320

[151] Diaz-Serrano A, Angulo B, Dominguez C, Pazo-Cid R, Salud A, Jimenez-Fonseca P, et al. Genomic profiling of HER2-positive gastric cancer: PI3K/Akt/mTOR pathway as predictor of outcomes in HER2-positive advanced gastric cancer treated with trastuzumab. The Oncologist. 2018;**23**(9):1092-1102

[152] Fang WL, Huang KH, Lan YT, Lin CH, Chang SC, Chen MH, et al. Mutations in PI3K/AKT pathway genes and amplifications of PIK3CA are associated with patterns of recurrence in gastric cancers. Oncotarget. 2016;**7**(5):6201-6220

[153] Ito C, Nishizuka SS, Ishida K, Uesugi N, Sugai T, Tamura G, et al. Analysis of PIK3CA mutations and PI3K pathway proteins in advanced gastric cancer. The Journal of Surgical Research. 2017;**212**:195-204

[154] Owonikoko TK, Harvey RD, Carthon B, Chen Z, Lewis C, Collins H, et al. A phase I study of safety, pharmacokinetics, and pharmacodynamics of concurrent everolimus and buparlisib treatment in advanced solid tumors. Clinical Cancer Research. 2020;**26**(11):2497-2505

[155] Rodon J, Brana I, Siu LL, De Jonge MJ, Homji N, Mills D, et al. Phase I dose-escalation and -expansion study of buparlisib (BKM120), an oral pan-class I PI3K inhibitor, in patients with advanced solid tumors. Investigational New Drugs. 2014;**32**(4):670-681

[156] Kim KJ, Kim JW, Sung JH, Suh KJ, Lee JY, Kim SH, et al. PI3K-targeting strategy using alpelisib to enhance the antitumor effect of paclitaxel in human gastric cancer. Scientific Reports. 2020;**10**(1):12308

[157] Jung EJ, Lee YM, Lee BL, Chang MS, Kim WH. Lytic induction and apoptosis of Epstein-Barr virus-associated gastric cancer cell line with epigenetic modifiers and ganciclovir. Cancer Letters. 2007;**247**(1):77-83

[158] Chang MS, Uozaki H, Chong JM, Ushiku T, Sakuma K, Ishikawa S, et al. CpG island methylation status in gastric carcinoma with and without infection of Epstein-Barr virus. Clinical Cancer Research. 2006;**12**(10):2995-3002

[159] Schneider BJ, Shah MA, Klute K, Ocean A, Popa E, Altorki N, et al. Phase I study of epigenetic priming with azacitidine prior to standard neoadjuvant chemotherapy for patients with resectable gastric and esophageal adenocarcinoma: Evidence of tumor hypomethylation as an indicator of major histopathologic response. Clinical Cancer Research. 2017;**23**(11):2673-2680

[160] Kusano M, Toyota M, Suzuki H, Akino K, Aoki F, Fujita M, et al. Genetic, epigenetic, and clinicopathologic features of gastric carcinomas with the CpG island methylator phenotype and an association with Epstein-Barr virus. Cancer. 2006;**106**(7):1467-1479

Chapter 4

Ocular Manifestations in Epstein Barr Virus Infection

Andi Arus Victor

Abstract

The Epstein-Barr Virus (EBV), a member of the Herpesvirus family, occurs commonly and infects more than 90% of people worldwide. Most of systemic EBV infections throughout childhood and adulthood are typically asymptomatic or paucisymptomatic. Even though ocular involvements in EBV infections are infrequently reported, an increasing number of ocular manifestations have been previously reported . Ocular manifestation caused by EBV infection involved all segments of the eye, including oculoglandular syndrome, dry eye syndrome, dacryoadenitis, conjunctivitis, episcleritis, keratitis, uveitis, choroiditis, retinitis, retinal vasculitis, and papillitis. Previous reports found neurologic complications such as papilledema, optic neuritis, ophthalmoplegia, impaired accommodation, and facial nerve palsy. Any atypical ocular inflammatory process should be considered EBV infection in the differential diagnosis. The ocular manifestations of systemic EBV infection are varied and have not been emphasized. And the role of EBV infection in ocular manifestations should be fully described.

Keywords: Epstein-Barr virus, oculoglandular syndrome, dry eye syndrome, dacryoadenitis, conjunctivitis, episcleritis, keratitis, uveitis, choroiditis, retinitis, retinal vasculitis, papillitis, papilledema, optic neuritis, ophthalmoplegia, impaired accommodation, facial nerve palsy

1. Introduction

The Epstein-Barr Virus (EBV) was first introduced in 1964 by Epstein, Anchong, and Barr [1–3], from a tissue sample of Burkitt's lymphoma and was observed with an electron microscope. EBV has the same characteristic as herpes viruses in tissue cultures from specimens of Burkitt's lymphoma. EBV is included in the family of *Herpesviridae*, which has a characteristic of the viral genome covered inside a nucleocapsid bounded by the viral envelope. In 1968, 4 years right after its first discovery, EBV was found to be the causal agent of the infectious mononucleosis (also called glandular fever). On other studies that conducted after, EBV was also found on tissue samples of nasopharyngeal carcinoma [3, 4], non-Hodgkin's lymphoma, T-cell lymphoma, as well as oral hairy leukoplakia that associated with Acquired Immunodeficiency Syndrome (AIDS) [4, 5].

EBV DNA has a nature of double-strand and consists of guanosine and cytosine as much as 58%. Due to the nature of the receptors on B-cell surfaces, EBV shows a great permissivity toward B lymphocytes. When B lymphocytes are infected by EBV, they will be "transformed" into lymphoblasts in vitro, this process known as immortalization. EBV occurs commonly and infects approximately 90% of human [1–3].

EBV-specific antibodies are found in about 95% of adults and 50–85% of the children [4, 5]. EBV infections are typically asymptomatic or have a subclinical infection, but once infected, it can lead to a viral carrier state for a lifetime. In some cases, primary EBV infection occurring during early adulthood could show clinical significance that is known as glandular fever or infectious mononucleosis [2, 5, 6]. Fortunately, most patients with infectious mononucleosis show a short recovery period since both cellular and humoral immune responses play a prominent role during the infection.

Ocular manifestations caused by EBV infection have been previously described. Previous studies showed that specific antibodies were detectable in 65% of tear and 87.5% of serum samples from 40 normal subjects [4]. EBV infection manifested in the eye may involves all segments of the eye, including oculoglandular syndrome, dry eye syndrome, dacryoadenitis, conjunctivitis, episcleritis, keratitis, uveitis, choroiditis, retinitis, retinal vasculitis, and papillitis [3]. Follicular conjunctivitis is the most common ocular disease during acute EBV infection. It is observed in 1–38% of cases and is usually unilateral. Bilateral stromal keratitis with coin-shaped lesions noted on tapering of systemic steroid therapy has been reported [6]. Uveitis occurred in 0.5% of patients with EBV infection [6].

The ocular manifestations of systemic EBV infection are numerous and have not been emphasized. Therefore, the role of EBV infection in ocular manifestation should be fully understood.

2. Pathophysiology infection of EBV

Usually, the first introduction to EBV in the period of childhood will cause subclinical infection, however if this primary event happens during adolescence, most of the time would trigger infectious mononucleosis. This syndrome consists in an acute and self-limiting lymphoproliferation of infected B cells and, at the same time, the development of virus-specific T cells that are triggered in order to tackle viral dissemination.

EBV transmission occurs by the sharing of saliva and then cause infection of mucosal surfaces and lymphoid tissues [3]. Almost most of seropositive individuals shed virus in their saliva. It was proposed earlier that initial replication of the virus happened also in the epithelial cells of the oropharynx, which causes the B cell infection by the previously infected epithelial cells. However, newer studies suggest that B cells located in the oropharynx could be infected first and act as the primary site.

During the latent stage of the infection, the site in which EBV persists within the body is the resting memory B cells. The viral replication in the oropharynx is partially suppressed in patients taking acyclovir, in the other hand, the amount of B cell EBV infected in the circulation remains the same.

Typical symptoms and signs are fever, pharyngitis, lymphadenopathy, and splenomegaly. The basic lesion is a perivascular infiltration of both normal and abnormal lymphocytes in all the tissues except bone marrow [7].

These cells are metaplastic noninvasive and to be formed *in situ* from other cells of the reticuloendothelial system. This lesion distributed throughout the body, and in any individual patient may be most marked in the central nervous system, liver, lungs, and other organs or systems.

Before the entry process into B cell, gp350, a major envelope glycoprotein, fixes to the receptor of the virus, CD21 molecule, which is located on the surface of the B cell. There are other factors besides CD21 that play a significant role, such as major histocompatibility complex (MHC) class II molecules that acts as a cofactor during B cells infection. EBV genome linear DNA molecules encode 100 viral proteins.

These proteins are significant in the process of construction of virion as well as carrying forward immune responses of the host.

EBV infection in vivo of the epithelial cells consequences in active replication while simultaneously with lysis of the cells as well as production of virus. Meanwhile, EBV infected B cells in vitro will become immortalized and causes latent infection. Only a small portion of cases that viral replication is triggered immediately after the infection of B cells.

Both cellular and humoral immunity plays an important role in the infection of EBV in humans. Even though the presentation of antibodies designed against viral structural proteins is critical, cellular immunity plays a higher role on the controller of EBV infection. The control of proliferation of EBV infected B cells in the primary infection is carried forward by CD4+, CD8+, and natural killer cells. During the period of infectious mononucleosis, a proportion of 40% of CD8+ T cells are focused on single replicative EBV protein sequence, meanwhile only 2% are focused to single latent EBV protein sequence.

During acute infection, first IgM and then IgG antibodies to viral capsid antigens (VCA) appear. Anti-VCA IgG may lead to virus carrier state and persist for a lifetime [3, 7]. Antibodies to early antigens rise during the acute phases of the disease and subsequently decrease to low or undetectable.

EBV can exist regardless of competent responses from the healthy immune system that are targeted toward it. It would most probably suggest that EBV has developed some tactics to evade the response from the immune system. EBV encrypts a cytokine as well as cytokine receptors that have a crucial role in moderating capable immune system therefore this allows tenacious infection. The 70% of amino acid sequence of EBV BCRF1 has the same characteristics as interleukin-10.

The protein of BCRF1 imitates the behavior of interleukin-10 of inhibits interferon-γ synthesis by in vitro human peripheral-blood mononuclear. In addition to that, EBV BARF1 protein acts as a soluble receptor for colony-stimulating factor 1. It is known that colony- stimulating factor 1 usually upregulates the expression of interferon-α of monocytes, therefore BARF1 protein able to operate as a decoy receptor to halt the action of cytokine. Since inhibition of the outgrowth of EBV infected cells in vitro is done by interferon-γ and interferon-α, BARF1 and BCRF1 proteins could aid the EBV to avoid the immune system of the host in the period of both acute as well as latent infection. EBV also codes more than two proteins that cause apoptosis inhibition. Protein of the EBV BHFR1 is similar to the protein of human bcl-2 protein, this protein has the same function on blocking apoptosis. In addition, EBV LMP-1 increases several cellular protein expressions that hinders the process of apoptosis, i.e. bcl-2.

Antibodies to EBV nuclear antigens appear weeks to months later, providing serologic evidence of past infection. Lesions of the special sense organs are somewhat rare, but there are few reports of eye lesions or manifestations [7].

Ocular manifestations have been reported to affect all segments of the eye and most commonly associated with acute mononucleosis [3]. Manifestations in the ocular may be the cause of direct involvement of the eye and its adnexa through inflammatory syndrome similar to that of infectious mononucleosis, and those affecting vision and the neuro-ophthalmologic apparatus owing to a more remote occurrence of the lesion, most commonly involving the central nervous system [7].

3. Clinical manifestation

Ocular manifestations of acute EBV infection may affect the central nervous system to the extent of disturbance in the visual or oculomotor pathways, in addition it could as well cause a disorder in the eye and adnexa of the eye [4, 8]. Further manifestations were observed due to the advancement of diagnostic tools that could give much more sensitive and specific results [8].

Neurologic abnormalities caused by EBV could disturb the host vision. Disorders that have been reported previously consisted of papilledema, convergence deficiency, nerve palsies, retinal necrosis, central nervous system vasculitis, retinochoroiditis, necrotizing retinitis with extensive hemorrhage that were co-infected with human immunodeficiency virus and cytomegalovirus and optic neuritis [9].

The appearance of yellowish, coalescing lesion in the macula as well as edematous optic disk were seen in a patient with EBV infection by Kim et al. [10]. From our previous case-report, it was examined from funduscopy a peculiar finding of prominent presentation of white sheathing retinal phlebitis covering all four retinal quadrants (**Figure 1**). Other reports had described this finding as frosted branch angiitis. It was also well known that frosted branch angiitis has been seen regularly in numerous disorders such as Crohn disease, Behcet disease, and systemic lupus erythematosus. However, from our previous case, the patient did not show any typical symptoms of Crohn disease, Behcet disease, and systemic lupus erythematosus, also, PCR was done on the patient's serum and was positive for EBV.

EBV infects lymphoid tissues and also mucosal surfaces. In the cellular level, EBV infects B cell and conforming virus-specific T cells, which will lead to symptoms such as lymphadenopathy and pharyngitis. Based on seroepidemiologic data, it was reported that EBV infects all parts of the eye, though the most common abnormality includes follicular conjunctivitis. After the availability of serologic virus-specific tests, other disorders of the anterior segment that is associated with EBV were detected more frequently, which includes keratitis, iritis, episcleritis, as well as dacryoadenitis.

It was found by Plugfelder et al., [11], that some cases of primary Sjogren's syndrome developed right after infectious mononucleosis. These authors [11] gathered patients that have aqueous tear deficiency and evaluated them for a serologic finding of EBV infection. A significant correlation was found between elevated early antigen EBV titers and severe aqueous tear deficiency once the results were placed in multivariate analysis. This suggests that EBV infection could be a risk factor in developing aqueous tear deficiency [11, 12].

Inflammation of the conjunctiva has been seen and linked with keratitis, that variate from mild hyperemia to mild follicular reaction in the inferior and superior

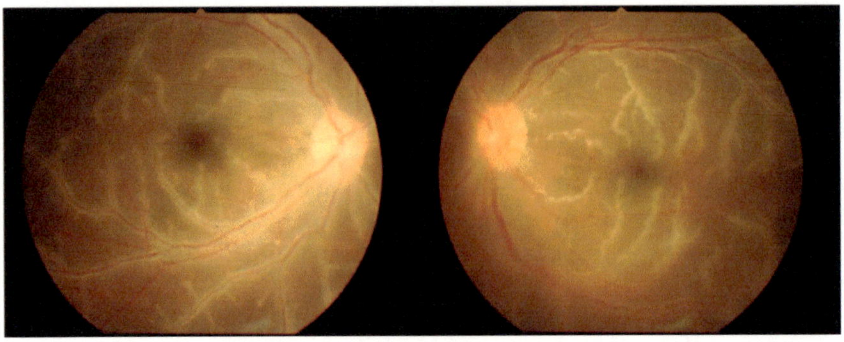

Figure 1.
Bilateral fundus photographs showing white sheathing of retinal veins in four retinal quadrants (frosted branch-like appearance) with macular edema. Reprinted from [1]. Copyright 2016 by Retinal Cases & Brief Reports.

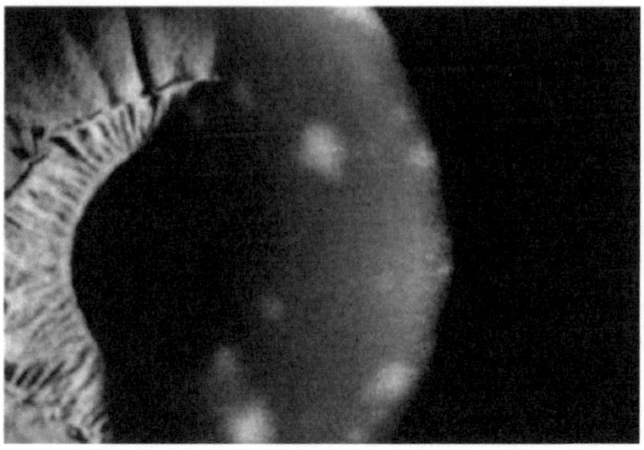

Figure 2.
Subepithelial infiltrates resembling adenoviral infection-associated keratitis in a patient with recent primary EBV infection. Reprinted from [2]. Copyright 1990 by Surv Ophthalmol.

tarsal conjunctiva. A study by Matoba [4] has found stromal keratitis related to EBV infection to infect all layers of stroma. Matoba [4] has also reported there were two forms of EBV-related keratitis, the first type was a granular, multiple, well-defined, ring, or circular-shaped opacities which was distributed all over the anterior as well as mid-stroma. Another form is nonsuppurative, multifocal keratitis including deep layers or full thickness of peripheral cornea and also related to various grades of vascularization [4]. In another study by Chodosh J, stromal keratitis associated with EBV infection presented three principal forms (**Figure 2**) [7].

Type I: multifocal subepithelial infiltrates that resemble adenoviral keratitis.

Type II: multifocal, blotchy, pleomorphic infiltrates with active inflammation or granular ring-shaped opacities (inactive form) in anterior to midstroma.

Type III: multifocal deep or full thickness peripheral infiltrates, with or without vascularization.

Matoba et al. [4] proposed a hypothesis about the pathophysiology of EBV related keratitis onset being carried forward by infectious and immunologic processes. Due to the quick response to the introduction of topical corticosteroid and seeming resolution of the inflammation without involving the use of antiviral therapy, it is more likely that immunologic processes were involved rather than viral replication. If this is true, the cornea was affected could be due to native keratocyte share a similar antigen with EBV, or it might be caused by EBV antigen develops located within the tissue [4].

Other common symptoms affecting the ocular include periorbital edema, pain when rotating the eyes, deep orbital pain, photophobia, as well as headache. It was spotted by Tanner [8] that episcleritis, as well as uveitis as part of the ocular manifestation of EBV infection. Tanner [8] reported nongranulomatous uveitis taking place at the end stage of clinical manifestation of infectious mononucleosis had been noted in four cases. One specific exhibition is the disorder of the oculomotor apparatus. Fledelius et al. [8] observed an exclusive inferior rectus paresis. Motto and Ashworth [8] were the first to notice a bilateral papilledema case of infectious mononucleosis with the nonappearance clinical manifestation of encephalitis nor meningitis. Other studies observed ptosis, nystagmus, as well as diplopia in their patients [8].

4. A diagnostic approach for detecting ocular disease by EBV infection

Most of the ocular manifestations linked to the EBV infection have been based mostly on seroepidemiologic data. This claimed is supported by the fact that numerous patients with EBV infection that showed ocular manifestations have been confirmed from in-situ hybridization of EBV genome from the biopsy of suspected tissue specimens.

The diagnosis of EBV infection by serologically for quite sometimes has been hinge on heterophile antibodies detection [4]. Heterophile antibodies are classified in the IgM group and would typically reach its highest levels around the second to third week since the first onset of the sickness and would exist and visible till 1 year [4]. They are detectable in up to 90% of adults who suffered from primary EBV infection [4].

Monospot test is a test that could be used in detecting EBV infection. This test is a quick slide agglutination test for specific heterophile antibodies produced by the human immune system in response to EBV infection. The sample will clump when it exposes to equine erythrocytes if these specific antibodies present in the patient's blood specimen. The Monospot test is considered to be a very specific test. However, its sensitivity falls in the range of 70–90% [13]. Patients with atypical clinical features and patients and with suspected of chronic infection are better evaluated with EBV specific serologic test that measure antibody levels against VCA, EBNA, and EA [4].

5. Management

The overall goal of EBV treatment is mainly supportive since the disease is usually self- limited [4]. Antiviral drugs have been used to inhibit the replication cycle of the virus. The action of antiviral agents can be divided into: (1) disrupting with cellular process which the virus uses for its replication; (2) inhibits the function of the virus by bind to the nucleic acid; (3) modifies the viral envelope which is resulting in preventing the virus infecting new cells; (4) inhibits the formation of new progeny by interfering with viral assembly; (5) interferes with the viral enzyme and inhibits their function; (6) prevents the processing of viral precursor polypeptide. The effect of some of the antiviral agents used in EBV infection is uncertain, but some studies have reported good results of treating EBV infection with systemic antiviral [4, 12].

Several antiviral drugs have been used to treat EBV infection and can be grouped into three as nucleoside analogs such as ganciclovir, valganciclovir, acyclovir, valaciclovir, acyclic nucleotide analog such as cidofovir and adefovir, and pyrophosphate analog such as foscarnet [12]. Acyclovir is recommended as the first-line drug for treating EBV infection. Therefore, it has become the most commonly prescribed antiviral regimen [14–16]. Anderson et al. [14] have found a significant reduction of EBV-infected B-lymphocytes after acyclovir treatment. Accordingly, systemic acyclovir therapy showed a good result of treating EBV-associated ocular involvement and Acute retinal necrosis [15].

Acyclovir triphosphate inhibits viral replication by acting as a competitive substrate for viral DNA polymerase, and its subsequent incorporation into the viral DNA chain results in obligate chain termination. The recommended regime is intravenous acyclovir 10 mg/kg every 8 hours (or 1500 mg/m^2) per day for 5–10 days, followed by oral acyclovir 400–800 mg 5 times daily for an additional 6–12 weeks. Second eye involvement may occur within the first 6 weeks after EBV infection; thus, the minimum duration of subsequent oral therapy was 6 weeks. Even though some ophthalmic centers are switching to oral therapy alone, few studies found a higher level of intravitreal acyclovir when given intravenously [15–17]. From the limited data available, one may conclude that acyclovir given intravenously and orally is the recommended regime [16, 17].

The effectiveness of steroid along with antiviral drug in treating ocular manifestations of EBV infection remains unclear. Some studies described that ophthalmic steroids relieve the symptoms of ocular inflammation in the anterior segment of the eye, including keratitis, anterior uveitis, and ocular allergies or injuries, yet its effect on the posterior part is still not clear [2, 4, 17].

6. Conclusion

It can be concluded that distinctive characteristic pathologic clinical manifestation, as well as lesions of infectious mononucleosis, exists in the ocular area. The ocular manifestations of systemic EBV infection are wide-ranging and have not been highlighted. EBV infection is best diagnosed using heterophile antibody tests to detect primary EBV infections and serology tests for asymptomatic patients. Despite of the self-limiting nature of the disease, acyclovir still becomes the first line of treatment with its attribute to reduce the viral replication. EBV infection should be included in the differential diagnosis of most unusual inflammatory course of the eye.

Conflict of interest

There are no conflicts of interest in this chapter.

Author details

Andi Arus Victor
Department of Ophthalmology, Faculty of Medicine, Cipto Mangunkusumo National General Hospital, Universitas Indonesia, Jakarta, Indonesia

*Address all correspondence to: arvimadao@yahoo.com

References

[1] Cohen JI. Epstein-Barr virus infection. The New England Journal of Medicine. 2000;**343**:481-492

[2] Victor AA, Sukmana N. Retinal vasculitis associated with Epstein-Barr virus infection, a case report. Retinal Cases & Bried Reports. 2016:1-4

[3] Slobod KS. Molecular evidence of ocular Epstein-Barr virus infection. Clinical Infectious Diseases. 2000;**31**:184-188

[4] Matoba AY. Ocular disease associated with Epstein-Barr virus infection. Survey of Ophthalmology. 1990;**35**:145-150

[5] Kramer S, Brumer C, Zierhut M. Epstein-Barr virus associated acute retinal necrosis. The British Journal of Ophthalmology. 2001;**85**:110-120

[6] Wong KW, D'Amico DJ, Hedges TR III, et al. Ocular involvement associated with chronic Epstein-Barr virus disease. Archives of Ophthalmology. 1987;**105**:788-792

[7] Chodosh J. Epstein-Barr virus stromal keratitis. Ophthalmology Clinics of North America. 1994;7(4):549-556

[8] Tanner OR. Ocular manifestations of infectious mononucleosis. In: Presented at the 36th Annual Meeting of the Pacific Coast Oto-Ophthalmological Society, 1st May 1952, Salt Lake City; 1952

[9] Andi AV. Ocular Manifestations in Epstein Barr Virus Infection [Online First]. IntechOpen; October 5th 2020.

[10] Kim SJ, Barananano DE, Grossniklaus HE, Martin DF. Epstein-Barr infection of the retina: Case report and review of the literature. Retinal Cases and Brief Reports. 2011;**5**:1-5

[11] Pfugfelder SC, Tseng SCG, Pepose JS, et al. Epstein-Barr virus infection and immunologic dysfunction in patients with aqueous tear deficiency. Ophthalmology. 1990;**97**:313-323

[12] Yaro A. Epstein-Barr infection: Current treatment options. Annals of Tropical Medicine and Public Health. 2013;**6**(1):10-13

[13] Stuempfig ND, Seroy J. Monospot test. [Updated 2019 Mar 10]. In: StatPearls [Internet]. Treasure Island (FL): StatPearls Publishing; 2020. p. 1

[14] Keorochana N. A case report of Epstein-Barr virus associated retinal vasculitis: Successful treatment using only acyclovir therapy. International Medical Case Reports Journal. 2016;**9**:213-218

[15] Rafailidis PI, Mavros MN, Kapaskelis A, Falagas ME. Antiviral treatment for severe EBV infections in apparently immunocompetent patients. Journal of Clinical Virology. 2010;**49**(3):151-157

[16] Patrick MK, Claire YH, Susan L. Antiviral selection in the management of acute retinal necrosis. Clinical Ophthalmology. 2010;**4**:11-20

[17] Lee FF, Foster CS. Pharmacotherapy of uveitis. Expert Opinion on Pharmacotherapy. 2010;**11**:1135-1146

Chapter 5

Epstein-Barr Virus in Myasthenia Gravis: Key Contributing Factor Linking Innate Immunity with B-Cell-Mediated Autoimmunity

Federica Bortone, Letizia Scandiffio, Paola Cavalcante, Renato Mantegazza and Pia Bernasconi

Abstract

Epstein-Barr virus (EBV), a common human herpes virus latently infecting most of the world's population with periodic reactivations, is the main environmental factor suspected to trigger and/or sustain autoimmunity by its ability to disrupt B-cell tolerance checkpoints. Myasthenia gravis (MG) is a prototypic autoimmune disorder, mostly caused by autoantibodies to acetylcholine receptor (AChR) of the neuromuscular junction, which cause muscle weakness and fatigability. Most patients display hyperplastic thymus, characterized by ectopic germinal center formation, chronic inflammation, exacerbated Toll-like receptor activation, and abnormal B-cell activation. After an overview on MG clinical features and intra-thymic pathogenesis, in the present chapter, we describe our main findings on EBV presence in MG thymuses, including hyperplastic and thymoma thymuses, in relationship with innate immunity activation and data from other autoimmune conditions. Our overall data strongly indicate a critical contribution of EBV to innate immune dysregulation and sustained B-cell-mediated autoimmune response in the pathological thymus of MG patients.

Keywords: autoimmunity, Epstein-Barr virus, innate immunity, myasthenia gravis, thymus, toll-like receptors

1. Introduction

Myasthenia gravis (MG) is a clinically heterogeneous, B-cell-mediated disorder affecting the neuromuscular junction (NMJ) and is mostly caused by the abnormal production of autoantibodies against the acetylcholine receptor (AChR) located in the postsynaptic membrane [1]. MG is generally considered a prototypical autoimmune disease, since the main target of the autoimmune response in the affected patients is well known. However, it is uniquely characterized by morphological and functional thymic abnormalities (hyperplasia and thymoma), which make this organ the main site of immunological alterations leading to the disease. Indeed, if the muscle is the target organ in MG patients, the thymus is now widely accepted as the main effector organ, in which B-cell expansion, anti-AChR autosensitization, and autoimmune response arise and are perpetuated [2]. The exact mechanisms triggering and sustaining autoimmunity in MG thymus are still unknown. As with many autoimmune conditions, considerable evidence indicates a multifactorial MG pathogenesis based on complex interactions among multiple genetic and environmental factors and their interplay with the immune system [3]. Among environmental factors, viruses are the main suspects to play a role in autoimmune diseases, mainly by their ability to induce persistent or aberrant Toll-like receptor (TLR)-mediated innate immune responses, which are able to promote or favor autoimmunity in genetically susceptible individuals. In our studies, we provided evidence of dysregulated Epstein-Barr virus (EBV) infection in hyperplastic and thymomatous MG thymuses in association with TLR overexpression, thus revealing EBV as a key contributing factor to intra-thymic B-cell tolerance disruption and sustained B-cell-mediated autoimmunity in MG patients [4].

2. Myasthenia gravis: autoantibodies and clinical features

MG is an autoimmune disorder of the NMJ, leading to fluctuating weakness and fatigability of skeletal muscles, exacerbated by repetitive contraction and improved on resting. Frequently, MG starts with ocular symptoms, as diplopia and ptosis, but in 80–85% of cases, ocular disease progresses to a generalized form within the first 2–5 years from onset, involving skeletal, bulbar, or respiratory muscles [5]. Respiratory failure (myasthenic crisis) occurs in 15–20% of patients and can be observed in younger and older patients [6].

MG is a heterogeneous condition whose clinical variability allows classification of patients in distinct disease subgroups, mainly based on autoantibodies, age at onset, and thymic histology [1, 7, 8]. In more than 80% of patients, the autoimmune attack is mediated by autoantibodies against AChR, less frequently (1–5%) against the muscle-specific kinase receptor (MuSK) or the low-density lipoprotein receptor-related protein 4 (LRP4), two proteins involved in AChR clustering. In addition, autoantibodies against other NMJ components, including cortactin, agrin, titin, and ryanodine receptor (RyR), have been described, especially in late-onset or thymoma-associated disease; their presence is concomitant to anti-AChR autoantibodies and indicates more severe manifestations [7]. Around 10% of generalized (non-ocular) patients results negative for anti-AChR, anti-MuSK, or anti-LRP4 antibodies. Clinically, these seronegative patients are similar to AChR-MG patients and can show thymic hyperplastic changes [9]. The triple seronegative MG subgroup may be heterogeneous, including patients with antibodies having affinities or concentrations too low to be detected with standard routine assays, or patients with antibodies against relevant antigens not identified yet. The introduction of cell-based assays (CBAs), having increased sensitivity compared to the routine assays commonly used, has significantly increasing the chance to identify autoantibodies to low-affinity clustered AChR, MuSK, and LRP4, thus improving MG diagnosis [7, 10–12].

As mentioned above, AChR-MG is associated with thymic patho-histological changes, including follicular hyperplasia and thymoma [13, 14]. Its severity is related with the loss of AChRs on NMJ, but not with the titers of circulating autoantibodies [15]. According with age at onset, AChR-MG follows a bimodal pattern, with the first peak under 50 years (early-onset MG, EOMG), and a second peak >50 years (late-onset MG, LOMG) [16, 17]. EOMG occurs most in young women and is generally associated with thymic hyperplasia. LOMG, which usually is generalized, mainly affects men, who frequently present thymic involution. Thymoma-associated MG that presents more severe symptoms can occur at any age, though it is more frequent in the elderly myasthenic patients and is frequently associated with the presence of antibodies against RyR and titin along with anti-AChR antibodies [14, 17].

MuSK-MG patients are mainly young females and typically have severe clinical symptoms [18]. An intra-thymic pathogenesis for MuSK-MG is not considered relevant, although a recent study found hyperplasia in 23% of MG patients positive for anti-MuSK antibodies by CBA [11]. LRP4-MG is less characterized but largely overlaps with AChR-MG clinical features; typical thymic histopathology has been recorded with a sparing of thymoma, at least in the European multinational cooperative study by Zisimopoulou and colleagues [12]. Most of LRP4-MG patients present ocular or generalized mild manifestations, and about 20% have only ocular weakness for more than 2 years [12, 19].

In MG, degradation of the postsynaptic membrane results in decreased AChRs and voltage-gated sodium channels, causing a significant reduction of endplate potential and raising the firing threshold, which is required to generate an action potential. Thus, during prolonged synaptic activity, as the quantal ACh content normally runs down, the summation of endplate potentials falls below the threshold, and they can no longer trigger the action potential of muscle fibers, leading to typical muscle weakness [20]. Three mechanisms of action of anti-AChR autoantibodies can explain NMJ impairment: (1) functional AChR block due to autoantibodies binding to the ACh-binding sites; (2) cross-linking and subsequent AChR internalization due to the ability of autoantibodies to target two antigen molecules (antigenic modulation); and (3) complement pathway activation, which leads to the generation of the membrane attack complex (MAC) and hence the destruction of the postsynaptic membrane. Complement activation at NMJ is thought to be the main pathogenic mechanism of anti-AChR antibodies, which are mainly of the IgG1 and IgG3 classes and therefore can bind and activate the complement system [20].

MuSK antibodies are generally IgG4, lacking complement-fixing, and are considered functionally monovalent, being unable to induce antigenic modulation. They affect MuSK ability to maintain the correct AChR cluster at the NMJ, by inhibiting the formation of MuSK-LRP4 complex and the agrin-stimulated MuSK phosphorylation [21, 22]. In addition, anti-MuSK antibodies are able to block binding of ColQ to the NMJ, compromising agrin-mediated AChR clustering [23]. Experimentally, animals receiving repeated daily injections of MuSK-positive patients' IgG, or actively immunized with MuSK, show impaired neuromuscular transmission, with reductions in endplate AChRs [24].

Both AChR-MG and MuSK-MG fulfill Witebsky's criteria for autoimmune diseases [25]. LRP4-MG also seems to adhere to these criteria: mice immunized with the LRP4 extracellular domain, or with IgGs purified from LRP4-immunized rabbits, present anti-LRP4 antibodies, exhibit MG-associated symptoms and their serum is able to decrease cell surface LRP4 levels [26, 27]. Anti-LRP4 autoantibodies mainly belong to the IgG1 subclass, thus they can activate the complement system; moreover, they prevent agrin-induced MuSK activation and AChR clustering [27].

Current therapeutic approaches for MG include symptomatic treatment with cholinesterase inhibitors, non-specific immunosuppression with corticosteroids and thymectomy in selected patients. Plasmapheresis or immunoglobulins are used for acute management of severe muscular weakness [8]. New biological drugs targeting molecules involved in the specific immune-pathological mechanisms, like eculizumab, which blocks the C5 terminal complement component, and efgartigimod, a functional IgG neonatal Fc receptor blocker, are promising for more specific and effective intervention to reduce corticosteroids side effects and to treat refractory patients [8].

3. Role of the thymus in MG pathology

The thymus is a primary lymphoid organ that provides a complex environment essential for T-cell differentiation and the establishment of central tolerance. It is composed of various cell types, mainly thymocytes and thymic epithelial cells (TECs), but also myoid cells, dendritic cells (DCs), macrophages, and B-cells in a limited number. By expressing tissue-specific antigens, mainly via the transcription factor autoimmune regulator (AIRE), medullary TECs play a key role in negative selection of thymocytes. Indeed, interactions between these cells and developing thymocytes lead to the elimination of autoreactive T cells, whereas the self-tolerant T cells continue their differentiation through the different thymus compartments, to be exported to the periphery [28].

Structural and functional pathological alterations of the thymus are found in approximately 80% of AChR-MG patients with generalized disease, including thymic hyperplasia (about 70% of patients) and thymoma (10–15%); the remaining patients (10–20%) present an atrophic or involuted thymus, mainly consisting of adipose tissue with residual areas of thymic parenchyma, in some cases showing hyperplastic changes (**Figure 1**) [13, 29, 30]. Hyperplasia is characterized by the

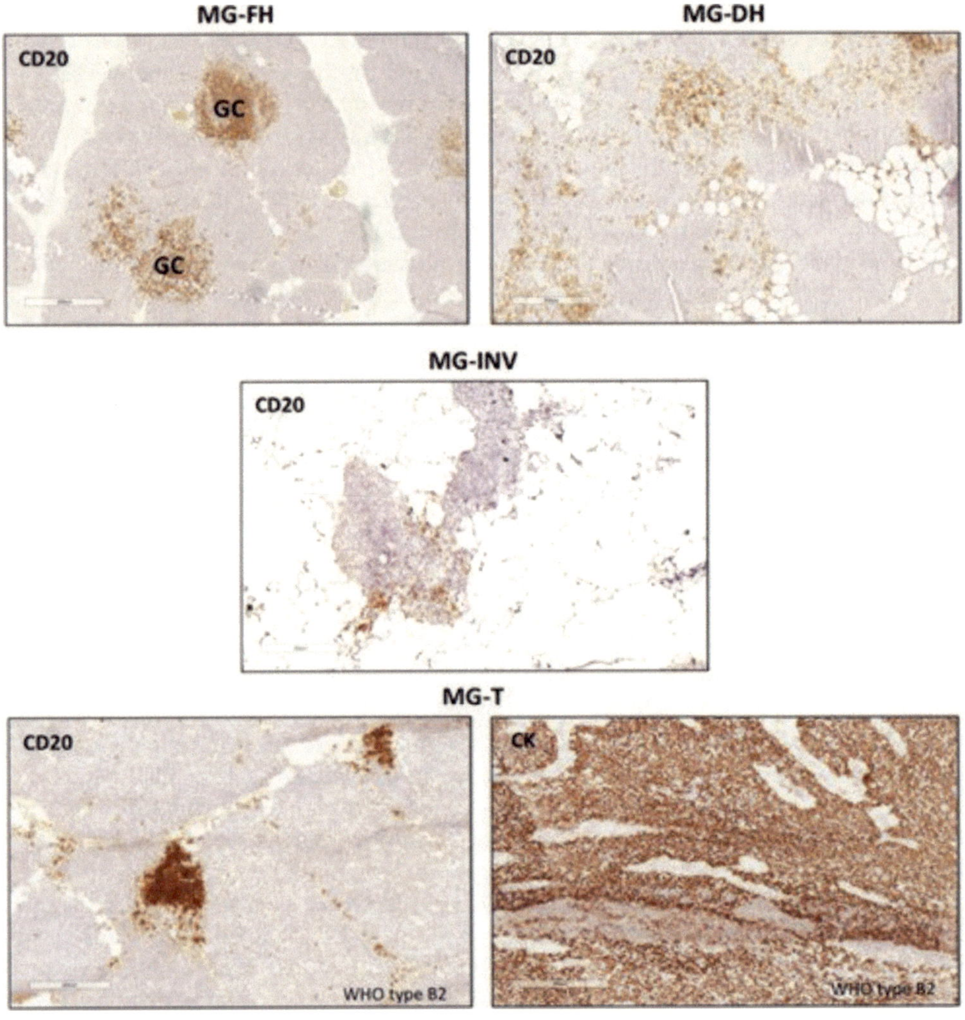

Figure 1.
Pathological abnormalities of myasthenia gravis thymus. Immunohistochemistry stainings showing CD20-positive B-cells in ectopic germinal centers (GCs) and lymphoid infiltrates of follicular hyperplastic (MG-FH) and diffuse hypeplastic (MG-DH) thymuses. CD20-positive B-cells are also scattered throughout the residual thymic parenchima in involuted MG thymuses characterized by abundant interlobular fat (MG-INV). CD20-positive B-cells, isolated or present as aggregates, can be detected in the cytokeratin (CK)-positive tumoral tissue of MG thymomas (MG-T).

presence of B-lymphocyte infiltrates invading the thymic medulla, or present in expanded perivascular spaces fused with the thymic medulla. B-cell infiltrates can be scattered throughout the medullary parenchyma (diffuse hyperplasia) or be organized into ectopic B-cell germinal centers (GCs) which, together with DCs and follicular helper T (Tfh) cells, form follicles (follicular hyperplasia) [30]. GCs are microarchitectures specialized to produce high-affinity antibodies against antigens, to establish the humoral immune response [31]. They are present in secondary lymphoid organs but rarely into the thymus, implying that GC development in the thymus of MG patients is a pathological event related to autoimmunity development. Ectopic GC formation is a characteristic feature of other organs target of chronic inflammation and autoimmunity, including multiple sclerosis (MS) brain and synovia of rheumatoid arthritis (RA) patients, indicating that lymphoid neogenesis plays a relevant role in the immune-pathological process of inflammatory autoimmune conditions [31–33]. In MG, GC formation is associated with the production of high endothelial venules expressing inflammatory chemokines, that abnormally recruit peripheral immune cells into the thymus, including the chemokine ligand 21 (CCL21), a key molecule that orchestrates thymic hyperplastic changes by promoting B-cell infiltration [34]. Uniquely, MG thymic GCs are surrounded by muscle-like myoid cells expressing AChR and other muscle antigens, along with plasma cells [35], thus supporting the idea that GCs may be the site of autosensitization and autoantibody production in the thymus of MG patients.

A wealth of data indicates that MG thymus contains all the elements necessary for developing and perpetuating an AChR-specific autoimmune response: TECs and muscle-like myoid cells expressing the autoantigen, professional antigen-presenting cells, AChR-specific autoreactive T cells and B-cells, and plasma cells producing autoantibodies [30]. Indeed, transplantation of MG thymic fragments to immunodeficient mice induces the formation of anti-AChR antibodies and their deposition at the skeletal muscle endplates [36].

As regard to the autoantigen presentation, TECs express major histocompatibility (MHC) class II complex and AChR subunits, including α, β, and γ subunits [37]. Thymic myoid cells express not only AChR subunits but also a functional AChR, whose fragments can be presented to T lymphocytes by cross-presentation via DCs, since myoid cells do not express MHC class II molecules [38, 39]. Cross-presentation may be favored by a persistent autoantibody and complement-mediated attack to myoid cells, which make the levels of autoantigen more available to DCs [40, 41]. Both TECs and myoid cells respond to pro-inflammatory cytokines by increasing the expression of AChR components, mainly the α subunit, which contains the main immunogenic region, thus suggesting that inflammation in the thymic microenvironment can results in enhanced autoantigen presentation and possible autosensitization against AChR [42]. Indeed, several lines of evidence indicate that MG thymus is in a state of chronic inflammation, characterized by an overexpression of pro-inflammatory cytokines and chemokines (IL-6, CCL19, CCL21, CXCL10, CXCL11, CXCL13, and RANTES) [43]. Among cytokines, type I interferons (IFNs) and IFN-induced genes are significantly up-regulated in hyperplastic MG thymuses and have been critically involved in driving thymic events that can lead to follicular hyperplastic changes, AChR overexpression and autosensitization. In particular, IFN-β was found to increase α-AChR-specific expression in TECs, along with the expression of inflammatory chemokines (e.g., CXCL13 and CCL21), and was able to recruit T and B-cells into the thymus, as well as B-cell activating factor (BAFF), which favors B-cell survival [44]. In hyperplastic MG thymus, no changes in the frequency of CD4+ and CD8+ T cells exported to the periphery was observed, but functional defects of regulatory T cells (Tregs) were demonstrated, along with resistance of conventional T cells to the Treg immunosuppressive function, thus indicating dysregulation of immunoregulatory mechanisms [45]. Altered Treg/T effector cell balance was associated with increased expression of pro-inflammatory cytokines by MG T cells, mainly IL-17, IFN-γ, IL-21, and tumor necrosis factor α (TNF-α) [45].

Thymoma is a rare thymic epithelial tumor, associated with autoimmune and paraneoplastic syndromes. The most common thymoma-associated autoimmune disorder is MG: up to 50% of thymoma cases may develop MG, whereas 10–15% of MG patients present thymoma [46]. Histologically, thymoma is a slow growing, locally invasive tumor consisting of transformed epithelial cells surrounded by maturing polyclonal T cells. The most recent World Health Organization (WHO) classification identified five types, A, AB, B1, B2, and B3, based on the nature of the cortical or medullary epithelial cells, and on the proportion of lymphocytes, with B2 and AB being the WHO types most frequently associated with MG [47, 48]. MG associated with thymoma usually has worse prognosis, showing generalized and maximum severe symptoms [48]. Alterations typical of the thymoma microenvironment may explain the development of AChR-specific autoimmunity in thymoma patients. They include: the lack of a functional medulla expressing AIRE; the reduction, or absence, of tolerogenic myoid cells; the reduced expression of HLA class II molecules; and the failure of Treg generation, which ultimately leads to autosensitization to AChR, and other locally expressed muscle antigens, and defective negative T-cell selection [14, 49, 50]. Recently, a higher proportion of GCs was abnormally found in non-neoplastic thymic tissue adjacent to thymoma in MG patients compared to thymoma patients without MG, suggesting that B-cell dysregulation characterizes not only thymic hyperplasia but also thymoma-associated MG, and that GCs may represent a risk factor for the development of MG in thymoma patients [51].

An important evidence of the central contribution of the thymus in autoimmune response development and maintenance in MG is the ability of thymectomy to improve the disease course in non-thymomatous patients over a 2-year period after the surgery, as demonstrated by the MGTX clinical trial, and its extension study [29, 52]. Thymectomy is mandatory for thymoma patients. Its efficacy as treatment option in non-thymomatous patients is plausibly due to eradication of the site of autoimmunity, as indicated by the fact that AChR antibody titers usually fall after thymectomy, and the magnitude of this fall correlates with the proportion of GC B-cells in the removed thymus [53]. However, autoantibody titers do not always fall, suggesting other possible sites of autoantibody production in some patients [54].

3.1 Innate immunity and toll-like receptors: a pathogenic link with autoimmunity

Innate immunity is the first line of defense against pathogen infections. Its interplay with the adaptive-humoral immune system plays a key role in central and peripheral tolerance maintenance, and strictly depends on a fine regulation of TLRs. TLRs are a family of pattern recognition receptors (PRRs) able to recognize specific conserved microbial-derived molecular structures, thus sensing danger signals. They are expressed by a variety of cell types, but mainly by innate immune cells, such as DCs and macrophages [55]. The TLR family includes at least 11 members in humans: TLR1, TLR2, TLR4, TLR5, TLR6, and TLR10, which are located on the cell surface and recognize microbial membrane-associated molecules (e.g., LPS, lipoprotein, and peptidoglycan); TLR3, TLR7, TLR8, TLR9, and TLR11, present on the intra-cellular endosome membranes and able to distinguish bacterial or viral nucleic acids, including ssRNA, dsRNA, and unmethylated cytosine phosphate guanine (CpG)-containing DNA [55]. Upon its own specific ligand binding, TLR dimerizes, internalizes, and interacts with the intracellular adaptor myeloid differentiation primary response gene 88 (Myd88) or with the TIR domain-containing adapter-inducing IFN-β (TRIF), which activates nuclear factor κB (NFκB) and IFN response factor transcription. The TLR signaling cascade induces the expression of pro-inflammatory agents (e.g., IFN-I, IL-6, IL-12, IL-23, and TNF-α), which in turn contribute to activate immune system cells [56]. In this way, the first function of TLRs is to set up an innate immune response to protect the organism from pathogens. However, TLRs have been implicated in several autoimmune diseases, including systemic lupus erythematosus (SLE), RA, and MS, by studies performed in humans and animal models [56, 57]. Specifically, dysregulated or persistent TLR activation has been demonstrated to contribute to autoimmunity by (i) abnormal stimulation of antigen-presenting cell maturation and increased IFN-I and pro-inflammatory cytokine production; (ii) altered balance between Treg and T helper 17 (Th17) cells; (iii) induction of co-stimulatory signals for proliferation, maturation, and survival of B-cells, which compromise B-cell tolerance; and (iv) promotion of GC formation [56–58].

In hyperplastic MG thymuses, overexpression of TLR3, TLR4, TLR7, and TLR9, in association with chronic inflammation, has been demonstrated in different studies, thus supporting the existence of a critical cross talk between innate immunity and autoimmunity in the intra-thymic MG pathogenesis [59–63].

Our group was the first to demonstrate a marked overexpression of TLR4 in involuted and hyperplastic MG thymuses, especially in TECs [59]. Later, we revealed that TLR4 stimulation in MG TECs was able to increase the production of Th17-related cytokines and the expression of CCL17 and CCL22, two chemokines involved in peripheral immune system cell recruitment in inflamed organs, that we found to be overexpressed in MG thymuses [60]. Moreover, by generating an in vitro imaging model based on experimental autoimmune MG co-cultures of Th1/Th17 AChR-specific T cells, naïve Tregs, DCs, and TECs, we found that TLR4 stimulation increased AChR-specific T-cell activation and impaired Treg function, thus disclosing a contribution of dysregulated TLR4 signaling to the inflammatory and autoimmune process in the MG thymic milieu [60].

Cufi and colleagues showed that also TLR3 is overexpressed in MG thymuses and demonstrated that stimulation of this TLR with its ligand poly(I:C), a synthetic analog of viral dsRNA, induced a specific up-regulation of α-AChR in TECs but not of other AChR subunits or tissue-specific antigens, via IFN-β release [61]. Of interest, another study of the same group disclosed that poly(I:C) injection in wild-type mice, in combination with LPS, that stimulates TLR4 was able to increase α-AChR thymic expression and induce thymic hyperplastic changes along with production of serum anti-AChR antibodies [62]. These mice developed MG symptoms in absence of any AChR immunization, thus supporting the idea that pathogen infections could contribute to anti-AChR sensitization and autoimmunity development via persistent or abnormal TLR stimulation [62].

Along with TLR3 and TLR4, TLR7 and TLR9 have also been implicated in the intra-thymic pathological events leading to MG. Indeed, we recently revealed a significant TLR7 and TLR9 up-regulation in both involuted and hyperplastic MG thymuses compared to normal thymuses, with the two receptors being mainly expressed in B-cells, TECs, plasmacytoid DCs, and macrophages [63]. TLR7 was also enhanced in thymic myeloid DCs and its transcriptional levels positively correlated with those of IFN-β [63]. Interestingly, as described in the following paragraphs, the two receptors were markedly expressed in B-cells and plasma cells positive for EBV proteins, indicating EBV as contributing environmental factor implicated in dysregulated TLR activation in MG thymuses.

Of interest, due to the key contribution of TLRs to chronic inflammation and immune system dysregulation, their pathways are rapidly emerging as attractive targets for therapeutic strategies able to mitigate or inhibit autoimmune processes [64].

4. EBV as contributing factor to autoimmunity

EBV is the main virus suspected to play a role in autoimmune diseases due to its unique ability to infect, activate, and immortalize B-cells, allowing them to evade immune surveillance, at the same time, promoting inflammatory state via TLR-mediated innate immune responses [65, 66].

EBV is a DNA virus of the herpes virus family transmitted through saliva exchange and infecting approximately 95% of the world's population [65]. EBV primary infection mostly occurs during childhood and shows mild symptoms or more frequently none. However, in adolescence or adulthood, EBV causes infectious mononucleosis in 30–70% of cases, with up to 20% of B-cells being infected [67]. After resolution of primary infection, EBV persists lifelong in the host in rare circulating memory B-cells. In the latent state, it is not detectable by immune system and its genome circularizes and replicates together with the host's chromosomal DNA, resulting in a restricted expression of a maximum of nine viral genes: the EBV nuclear antigens (EBNA1, −2, −3A, -3B, and -3C), the leader protein (LP), and the latent membrane proteins (LMP1, −2A, and -2B). Different expression patterns of EBV latent genes determine the occurrence of EBV latency types I, II, or III, each type being associated with distinct EBV-related diseases [65, 68–70]. EBV-encoded small nuclear RNA (EBER) 1 and 2 as well as EBNA1 are expressed in all the latency types [65]. EBNA2 is expressed during latency type III (known as growth program), typical of newly infected naïve B-cells, lymphoproliferative disorders, and mononucleosis [68–71]. LMP1 and LMP2A, key proteins that rescue infected B-cells from apoptosis, act as functional homologs of CD40 and B-cell receptor (BCR) and are expressed during latency III and II (known as default program), with type II being observed in memory B-cells and GC cells, Hodgkin and non-Hodgkin lymphoma, and nasopharyngeal carcinoma [65, 68]. Finally, latency I (known as true latency) is characterized by EBNA1 and EBER expression only; it is typically observed in rare peripheral blood memory B-cells and is associated with Burkitt's lymphoma [65, 68].

The exact mechanism waking up lytic EBV activities is not clear, but it seems to be the result of dynamic interactions between the host's immune response and the infection state. When infected B-cells differentiate into plasma cells, the promoter of early lytic genes can be reactivated, driving expression of numerous proteins involved in viral activities (i.e., BSLF1, BALF2, BBLF4, and BALF5) [65]. Two genes, BZLF1 and BRLF1, which encode viral transcription factors, orchestrate the transition from viral latency to lytic infection [69]. New virions primarily infect B-cells, and the viral entry is mediated by viral gp350 protein binding to CD21 [70].

Uniquely, EBV ensures early mechanisms of immune evasion, such as the inhibition of IFN pathways, through a viral IL-10 homolog, the suppression of cytotoxic T-cell responses and the down-regulation of MHC class I and II expression. Moreover, some of the viral proteins are anti-apoptotic, including the early antigen restricted (EA/R), which is a Bcl2 viral homolog that protects infected B-cells from apoptosis and immortalize them [71]. Several mature EBV miRNAs also contribute to immune system alterations in the host by modulating expression of genes involved in immune recognition, antigen presentation and cellular migration, such as miR-BHRF1–3, which regulates IFN-inducible T-cell-attracting chemokine CXCL-11 expression, or miR-BART20-5p and miR-BART8, which affect the IFN-γ signal transduction pathway [72].

Despite it is innocuous in most cases, EBV has true pathogenic potential due to persistent latent infection with periodic reactivations. Disruption of the virus-host balance in susceptible individuals can favor autoimmunity or the abovementioned B-cell malignancies [68, 73]. Among autoimmune diseases, EBV has been associated with MS, SLE, and RA by a number of sero-epidemiological and immunological studies [65, 68, 73]. Moreover, EBV persistence and reactivation have been demonstrated in ectopic B-cell follicles detected in MS brains [74], in the Sjögren's Syndrome salivary glands [75], and in synovia from RA patients [76], suggesting that EBV might be a common pathogenic feature of autoimmune conditions characterized by B-cell activation and lymphoid neogenesis.

Several mechanisms have been described to explain how viruses may rise an autoimmune response, including molecular mimicry and immune system general activation via TLRs. Molecular mimicry is due to T-cell Receptor (TCR) and BCR recognition flexibility, so that a microbial peptides, structurally similar to a self-peptide, may trigger the autoimmune response [77]. As regard to EBV, EBV-encoded proteins, such as BZLF1, share regions with the host transcription factors of the fos/jun family and host ankyrin proteins, which anchor the cytoskeleton and regulate host transcription factors, including NF-kB, which is critically involved in the immune response [78]. Quantitative and qualitative differences in CD4+ T-cell response to EBNA1 have been described in MS patients [79], and a small percentage of EBNA1-specific T-cell clones cross-recognize myelin-derived epitopes [80]. A further well-characterized example of molecular mimicry between EBV and host proteins is the similarity of regions of EBNA1 with ribonuclear protein Smith (Sm) antigen or the Ro self-protein in SLE [81]. There are also several examples of molecular mimicry relevant for RA, related to aminoacid motif sharing between HLA-DRB1 and EBV gp110, cytokeratin and type II collagen, and citrullinated human fibrin and a citrullinated EBNA1 form [82].

Another hypothesis of virus-induced autoimmunity is bystander activation, in which viruses act as super-antigens that promote general activation of the immune system; in this context, specialized antigen presenting cells present self-antigens, obtained by inflamed tissue destruction or by the uptake of local dying cells, to autoreactive T cells [77]. In MS, bystander activation may result in central nervous system damage by CD8+ T-cell cytotoxicity and EBER-induced IFN-α production via innate immune response, both described in post-mortem brain tissue from MS patients [74, 83]. EBV can promote inflammation in the infected organs by innate immune system activation, through release of molecules able to activate diverse TLRs and type I IFN production [66, 84]. The EBV envelope protein gp350 stimulates TLR2, whereas EBERs bind TLR3; moreover, EBV RNA and DNA can activate pathways mediated by TLR7 and TLR9 [66]. Thus, EBV ability to stimulate persistent TLR signaling may also contribute to disrupt immune system balance, favoring chronic inflammation and autoimmunity.

5. EBV in the intra-thymic MG pathogenesis

Evidence of chronic inflammation, GC formation, and TLR overexpression in MG thymus, along with data demonstrating a role of TLRs in inducing anti-AChR autoimmunity and MG symptoms in mice, pointed out the idea that MG pathogenesis could be associated with viral infections [43, 59–63].

Based on the hypothesis that active EBV infection may be a common pathogenic event of organs site of autoimmunity, in our previous studies we searched for the expression of EBV markers in the thymus of MG patients. Of interest, we provided the first demonstration of EBV presence in both hyperplastic and thymoma MG thymuses, but not in normal thymuses and non-MG thymomas, obtaining results indicative of a contribution of the virus to B-cell dysregulation, TLR overexpression and B-cell-mediated autoimmunity in MG [4, 63, 85–87].

Earlier serological studies to associate MG with EBV produced contrasting results. One of them underlined no significant difference in incidence or antibody titers to EBV, cytomegalovirus, herpes simplex type 1 and other virus, in 104 MG patients compared to age-matched healthy controls, weakening the virus hypothesis in MG pathogenesis [88]. More recently, Bhibhatbhan and colleagues reported a case of young woman with abrupt onset of both MuSK-MG and type I diabetes mellitus, following infectious mononucleosis, bringing attention back to EBV in MG [89]. Few years after, according with a possible association between MG and EBV, Csuka and collaborators discovered a strong association between EOMG and high anti-EBNA1 IgG serum concentration in EOMG patients [90]. Despite these results were conflicting, our recent pathological findings, which are deeply described below, allowed us to include MG among the autoimmune diseases critically associated with EBV.

5.1 EBV in hyperplastic MG thymus

To our knowledge, the earliest study attesting EBV presence in MG thymuses was performed by McGuire and colleagues, who found EBV DNA in thymuses of 2/4 MG patients with thymic hyperplasia, and 2/2 patients with thymoma [91]. Before that, several attempts were made for identifying or isolating viruses from homogenates or cell suspensions of MG thymuses, but without success [92, 93]. However, these initial studies used techniques, and tissue storage methods, that cannot be considered sufficiently sensitive or optimal today.

Based on data showing EBV reactivation in intra-meningeal B-cell follicles in MS patients [74], our group decided to check for signs of EBV infection in MG thymuses (n = 17), both hyperplastic (follicular and diffuse) and involuted thymuses by using a combination of techniques, including in situ hybridization (ISH) to detect EBERs, immunohistochemistry (IHC) for latent and lytic EBV antigens, real-time PCR for viral DNA (LMP1 gene), and nested PCR for viral transcripts. As controls, normal thymuses (n = 6) from adult cardiopathic donors without autoimmune diseases were analyzed [85]. Interestingly, all MG, but not normal thymuses, resulted positive for EBV latency and lytic markers, strongly indicating EBV persistence and reactivation as a common feature of MG patients' thymuses (**Figure 2**). In details, by ISH, variable number of EBERs-positive cells was detected in medullary infiltrates of all the MG thymuses analyzed, particularly in GCs of hyperplastic MG thymuses. Accordingly, IHC results showed expression of EBV latency proteins EBNA2, LMP1, and LMP2A in cells scattered throughout the thymic medulla and in GCs. The early BFRF1 and BMRF1, and the late p160 and gp350/220 lytic phase EBV proteins were also found in most MG thymuses, but not in control tissues, indicating EBV reactivation. Double immunofluorescence and confocal microscopy then revealed that latently infected cells were diffuse infiltrating and GC B-cells, whereas plasma blasts, mainly located around GCs, were positive for the lytic markers [85]. By PCR approaches, latent EBNA1 and LMP2A transcripts, and the early BZLF1 lytic transcript, along with EBV DNA, were detected in MG thymuses, but not in normal thymuses, thus confirming the active EBV infection. Due to the EBV properties to activate and immortalize B-cells, our findings thus suggested a critical contribution of the virus to intra-thymic B-cell dysfunction and B-cell-mediated autoimmunity in MG patients [85].

In a following study, we extended the analysis of EBV to additional MG thymuses (n = 19). Real-time PCR for EBV DNA (*Bam*HI-W repeat region), latent (EBER1, EBNA1, LMP1) and lytic (BZLF1) transcripts, and IHC for LMP1 and BZLF1 proteins, confirmed an active EBV infection in the thymus of MG patients

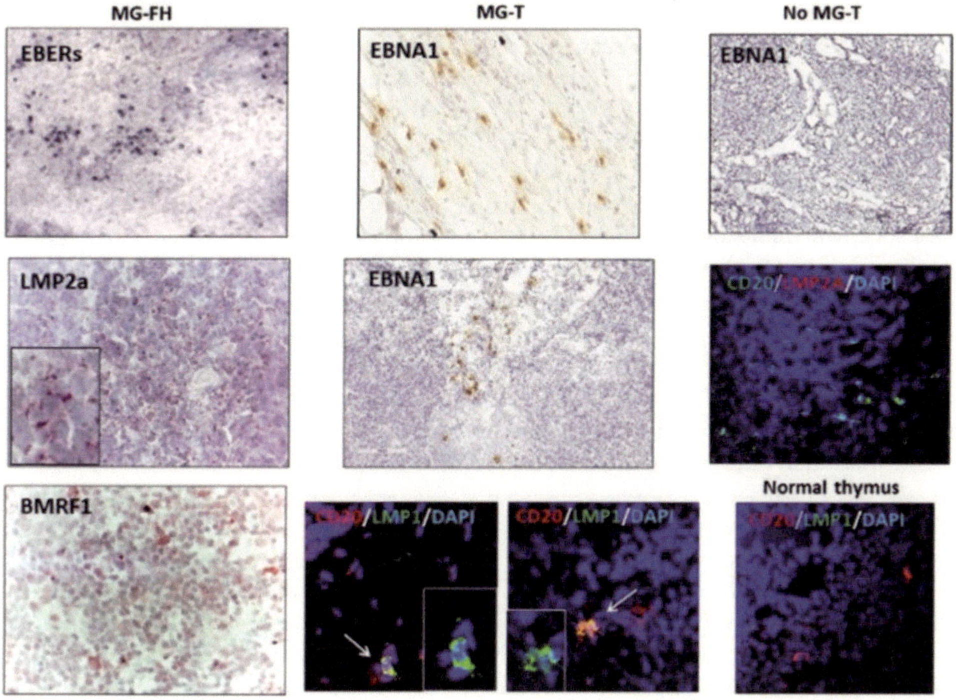

Figure 2.
Detection of EBV markers in follicular hyperplastic MG thymus (MG-FH), and MG thymoma (MG-T), but not in non-MG thymoma (No MG-T) and normal thymus. Representative images of immunohistochemistry and immunofluorescence stainings showing: EBER-positive cells and cells positive for latent LMP2A and lytic BMRF1 proteins in MG-FH; latent EBNA1-positive cells and CD20-positive B-cells expressing LMP1 in MG thymomas; absence of EBNA1-positive and LMP2A-positive cells in No MG-T; absence of LMP1-positive cells in normal thymuses.

but not in controls, thereby reinforcing the idea that the virus may significantly contribute to autoimmunity development or maintenance in MG thymus [86]. Similar results were found in MG thymuses from corticosteroid-naïve and -treated patients, indicating that the EBV infection profile we observed in our MG thymuses was not due to the immunosuppressive treatment [85, 86].

In contrast to our outcomes, Meyer and colleagues [94] and Kakalacheva and colleagues [95] reported absence or very low presence of EBV in MG thymuses. The first group performed ISH on formalin-fixed, paraffin-embedded MG thymic tissues (n = 44) and did not observe EBER-positive cells, likewise on cases of Hashimoto thyroiditis (n = 25), except for an isolate case in which rare EBER-positive lymphocytes were detected. Moreover, EBNA1 or BZLF1-positive cells were absent in MG thymuses or Hashimoto thyroiditis by IHC, whereas in infectious mononucleosis rare scattered positive cells were detected [94]. Kakalacheva *et al.* detected minimal levels of viral DNA in 6 of 16 hyperplastic MG thymuses, indicating rarity of viral copy numbers, confirmed by the observation of rare positivity for EBERs and viral proteins in the thymic sections by ISH and IHC [95]. Discrepancies between our data and data from these two studies may be due to sampling from patients with different clinical features, or to the use of different methods and tools with different sensitivity. In the article of Meyer et al., the cohort is apparently not homogeneous, and no description of treatment, nor clear indication of the patients' disease status, is given. Patients analyzed by Kakalacheva *et al.* did not require immunosuppression, so they maybe had a mild phenotype. As regards to the methods for EBV marker detection, tissue processing and procedures applied in these studies are different from ours, and perhaps less sensitive, or not optimized for detection of EBV in non-malignant pathological conditions [96]. All these issues need to be solved, to confirm data on EBV in MG thymus, and a correct approach could be to find an agreement on the best procedures to analyze EBV infection in the thymus.

To better understand the EBV role in MG pathogenesis, we recently investigated the potential cross talk between TLRs and EBV in rising or sustaining self-reactivity. In details, since EBV molecules (EBERs, EBV DNA) are known to activate TLR7 and TLR9, and considering that these two receptors have super-addictive effects on EBV-induced B-cell activation and transformation process [66, 97], we analyzed their expression in EBV-positive MG and EBV-negative normal thymuses. We revealed an increased percentage of proliferating B-cells positive for EBV markers, and overexpressing TLR7 and TLR9, in EBV-positive hyperplastic MG thymuses compared to controls [63]. Our overall data thus indicated for the first time that aberrant EBV-driven TLR7 and TLR9 signaling in MG thymuses might contribute to abnormal B-cell activation and proliferation, in turn promoting or perpetuating B-cell-mediated autoimmunity in MG patients.

5.2 EBV in MG thymoma

Since the 1980s, the involvement of EBV in thymoma, associated or not with MG, has long been investigated. However, data obtained by the different studies were contrasting: (i) McGuire et al. found EBV DNA in three thymomas, of which two were from MG patients [91]; (ii) absence of EBV in thymic epithelial tumors was reported by Inghirami *et al.* and Engel *et al.* [98, 99]; (iii) Chen et al. reported EBV DNA signals in eight out of 21 thymic carcinoma with lymphoepithelioma-like morphology, a subtype not currently included in the WHO classification, but they did not specify whether EBV-positive tumors were from MG patients [100]; and (iv) Takeuchi et al. detected EBV-infected lymphocytes in one out of 11 thymomas not associated with MG [101]. Recently, an antiviral gene signature was identified in MG thymomas by Cufi and colleagues, consisting in a significant overexpression of type I IFNs and components of the TLR3 signaling pathways [102], thus opening the hypothesis of a viral etiology also for MG associated with thymoma. The same authors tried to identify the presence of potential pathogens in the MG thymomas focusing on human papillomavirus (HPV) and EBV, but no viral DNA was detected [102].

Recently, by combining ISH, IHC and molecular techniques, our group demonstrated the presence of latency EBV markers in MG-associated thymomas, but only rarely in thymomas from patients without MG [87]. Specifically, by real-time PCR we showed a significantly higher frequency of EBV DNA and EBER1 detection in MG (53.8% and 84.6%) than non-MG (21.4%) thymomas, with the higher viral load and EBER1 levels being mainly observed in B2 and B2-mixed tumors, the WHO subtypes most frequently associated with MG. These data were in contrast with data on EBV DNA from Cufi and colleagues [102], likely because of the different sensitivity of the protocol used for the viral genome detection.

We then confirmed our molecular results by ISH, which showed the presence of cells positive for EBERs in MG, but not in non-MG thymomas [87]. Latent EBNA2 and late gp350/220 lytic transcripts were undetectable in all thymomas, except one, as well as the early lytic transcript BZLF1, thus revealing that early infection and EBV reactivation are very rare event in thymomas. By IHC analysis, we confirmed EBV persistence in MG thymomas, but not in thymomas without MG, and identified the phenotype of EBV-positive cells, that were B-cells positive for the EBV latency proteins EBNA1, LMP1, and LMP2, diffused throughout the MG neoplastic tissues (**Figure 2**). Based on the EBV gene expression pattern, we suggested EBV latency type II in MG thymomas. Due to the absence of EBV in thymic epithelial cells, these results thus revealed an association of EBV with B-cell-mediated autoimmunity in MG associated with thymoma, rather than with thymic neoplastic transformation [87]. We also found higher TLR3 transcriptional levels in MG than non-MG thymomas. Transcriptional levels of this receptor positively correlate with EBER1 levels, supporting a possible role for EBER1 in inducing persistent TLR3 signaling pathways in thymoma from MG patients [87]. Due to the previously described role of TLR3 in triggering an anti-AChR autoimmune response [62], our findings strengthened the idea of a critical contribution of EBV to B-cell-mediated autoimmunity via TLR3 in these pathological tissues. Indeed, the activation of EBV-driven TLR3 signaling may well contribute to create an altered tumor microenvironment, which supports the recruitment of peripheral B-cells, and thus B-cell dysregulation and tolerance disruption.

6. Conclusions

EBV has been associated with several autoimmune diseases by sero-epidemiological and immunological data. Our findings revealed for the first time a contribution of EBV to the intra-thymic MG pathogenesis, thus allowing to include MG among the autoimmune conditions associated with the virus. Based on our results and literature data, we postulate that, in the context of a genetically susceptible background, EBV persistence and reactivation into the thymus may favor B-cell dysregulation, TLR over-stimulation, inflammation, and B-cell-mediated autoimmune response to AChR, which can be perpetuated in the periphery. Further advancement in the knowledge of the exact involvement of EBV and TLRs in MG pathogenesis could set the basis for novel investigations aimed at developing innovative therapeutic applications targeting EBV-positive cells, TLRs, or components of their signaling pathways to counteract autoimmunity in MG and potentially other EBV-associated autoimmune diseases.

Acknowledgements

The authors wish to thank Dr. Francesca Aloisi and Dr. Barbara Serafini of the Department of Cell Biology and Neuroscience at the Istituto Superiore di Sanità (Rome, Italy) for the key contribution to the studies on EBV in MG thymuses. The authors also thank the neurologists and biologists of the Neurology IV Unit—Neuroimmunology and Neuromuscular Disease at the Fondazione I.R.C.C.S. Istituto Neurologico Carlo Besta (Milan, Italy), for their participation in various aspects of these studies, and the Italian Association of Myasthenia Gravis (A.I.M., Associazione Italiana Miastenia e Malattie Immunodegerative—Amici del Besta Onlus) for their kind support to the research activity. The research activity was supported by the Italian Ministry of Health (GR-2013- 02358564 and annual research funding).

Author details

Federica Bortone, Letizia Scandiffio, Paola Cavalcante*, Renato Mantegazza
and Pia Bernasconi
Neuroimmunology and Neuromuscular Diseases Unit, Fondazione I.R.C.C.S.
Istituto Neurologico Carlo Besta, Milan, Italy

*Address all correspondence to: paola.cavalcante@istituto-besta.it

References

[1] Mantegazza R, Bernasconi P, Cavalcante P. Myasthenia gravis: From autoantibodies to therapy. Current Opinion in Neurology. 2018;**31**(5):517-525. DOI: 10.1097/WCO.0000000000000596

[2] Berrih-Aknin S. Myasthenia gravis: Paradox versus paradigm in autoimmunity. Journal of Autoimmunity. 2014;**52**:1-28. DOI: 10.1016/j.jaut.2014.05.001

[3] Berrih-Aknin S, Le Panse R. Myasthenia gravis: A comprehensive review of immune dysregulation and etiological mechanisms. Journal of Autoimmunity. 2014;**52**:90-100. DOI: 10.1016/j.jaut.2013.12.011

[4] Cavalcante P, Barzago C, Baggi F, et al. Toll-like receptors 7 and 9 in myasthenia gravis thymus: Amplifiers of autoimmunity? Annals of the New York Academy of Sciences. 2018;**1413**(1):11-24. DOI: 10.1111/nyas.13534

[5] Hehir MK, Silvestri NJ. Generalized myasthenia gravis: Classification, clinical presentation, natural history, and epidemiology. Neurologic Clinics. 2018;**36**:253-260. DOI: 10.1016/j.ncl.2018.01.002

[6] Wendell LC, Levine JM. Myasthenic crisis. Neurohospitalist. 2011;**1**:16-22. DOI: 10.1177/1941875210382918

[7] Vincent A, Huda S, Cao M, et al. Serological and experimental studies in different forms of myasthenia gravis. Annals of the New York Academy of Sciences. 2018;**1413**:143-153. DOI: 10.1111/nyas.13592

[8] Mantegazza R, Cavalcante P. Diagnosis and treatment of myasthenia gravis. Current Opinion in Rheumatology. 2019;**31**(6):623-633. DOI: 10.1097/BOR.0000000000000647

[9] Gilhus NE, Skeie GO, Romi F, et al. Myasthenia gravis—Autoantibody characteristics and their implications for therapy. Nature Reviews. Neurology. 2016;**12**(5):259-268. DOI: 10.1038/nrneurol.2016.44

[10] Leite MI, Jacob S, Viegas S, et al. IgG1 antibodies to acetylcholine receptors in 'seronegative' myasthenia gravis. Brain. 2008;**131**(Pt7):1940-1952. DOI: 10.1093/brain/awn092

[11] Tsonis AI, Zisimopoulou P, Lazaridis K, et al. MuSK autoantibodies in myasthenia gravis detected by cell based assay—A multinational study. Journal of Neuroimmunology. 2015;**284**:10-17. DOI: 10.1016/j.jneuroim.2015.04.015

[12] Zisimopoulou P, Evangelakou P, Tzartos J, et al. A comprehensive analysis of the epidemiology and clinical characteristics of anti-LRP4 in myasthenia gravis. Journal of Autoimmunity. 2014;**52**:139-145. DOI: 10.1016/j.jaut.2013.12.004

[13] Cavalcante P, Cufi P, Mantegazza R, et al. Etiology of myasthenia gravis: Innate immunity signature in pathological thymus. Autoimmunity Reviews. 2013;**12**(9):863-874. DOI: 10.1016/j.autrev.2013.03.010

[14] Marx A, Porubsky S, Belharazem D, et al. Thymoma related myasthenia gravis in humans and potential animal models. Experimental Neurology. 2015;**270**:55-65. DOI: 10.1016/j.expneurol.2015.02.010

[15] Meriggioli MN, Sanders DB. Muscle autoantibodies in myasthenia gravis: Beyond diagnosis? Exp Rev Clin Immunol. 2012;**8**(5):427-438. DOI: 10.1586/eci.12.34

[16] Casetta I, Groppo E, De Gennaro R, et al. Myasthenia gravis: A changing pattern of incidence. Journal of

Neurology. 2010;**257**(12):2015-2019. DOI: 10.1007/s00415-010-5651-z

[17] Gilhus NE, Verschuuren JJ. Myasthenia gravis: Subgroup classification and therapeutic strategies. Lancet Neurology. 2015;**14**(10):1023-1036. DOI: 10.1016/S1474-4422(15)00145-3

[18] Evoli A, Alboini PE, Damato V, et al. Myasthenia gravis with antibodies to MuSK: An update. Annals of the New York Academy of Sciences. 2018;**1412**:82-89. DOI: 10.1111/nyas.13518

[19] Cordts I, Bodart N, Hartmann K, et al. Screening for lipoprotein receptor-related protein 4-, agrin-, and titin-antibodies and exploring the autoimmune spectrum in myasthenia gravis. Journal of Neurology. 2017;**264**(6):1193-1203. DOI: 10.1007/s00415-017-8514-z

[20] Conti-Fine BM, Milani M, Kaminski HJ. Myasthenia gravis: Past, present, and future. The Journal of Clinical Investigation. 2006;**116**(11):2843-2854. DOI: 10.1172/JCI29894

[21] Koneczny I, Stevens JA, De Rosa A, et al. IgG4 autoantibodies against muscle-specific kinase undergo fab-arm exchange in myasthenia gravis patients. Journal of Autoimmunity. 2017;**77**:104-115. DOI: 10.1016/j.jaut.2016.11.005

[22] Huijbers MG, Zhang W, Klooster R, et al. MuSK IgG4 autoantibodies cause myasthenia gravis by inhibiting binding between MuSK and Lrp4. Proceedings of the National Academy of Sciences of the United States of America. 2013;**110**(51):20783-20788. DOI: 10.1073/pnas.1313944110

[23] Kawakami Y, Ito M, Hirayama M, et al. Anti-MuSK autoantibodies block binding of collagen Q to MuSK. Neurology. 2011;**77**(20):1819-1826. DOI: 10.1212/WNL.0b013e318237f660

[24] Viegas S, Jacobson L, Waters P, et al. Passive and active immunization models of MuSK-Ab positive myasthenia: Electrophysiological evidence for pre and postsynaptic defects. Experimental Neurology. 2012;**234**(2):506-512. DOI: 10.1016/j.expneurol.2012.01.025

[25] Rose NR, Bona C. Defining criteria for autoimmune diseases (Witebsky's postulates revisited). Immunology Today. 1993;**14**(9):426-430. DOI: 10.1016/0167-5699(93)90244-F

[26] Mori S, Motohashi N, Takashima R, et al. Immunization of mice with LRP4 induces myasthenia similar to MuSK-associated myasthenia gravis. Experimental Neurology. 2017;**297**:158-167. DOI: 10.1016/j.expneurol.2017.08.006

[27] Shen C, Lu Y, Zhang B, et al. Antibodies against low-density lipoprotein receptor-related protein 4 induce myasthenia gravis. The Journal of Clinical Investigation. 2013;**123**(12):5190-5202. DOI: 10.1172/JCI66039

[28] Kurd N, Robey EA. T-cell selection in the thymus: A spatial and temporal perspective. Immunological Reviews. 2016;**271**(1):114-126. DOI: 10.1111/imr.12398

[29] Wolfe GI, Kaminski HJ, Aban IB, et al. Randomized trial of thymectomy in myasthenia gravis. The New England Journal of Medicine. 2016;**375**:511-522. DOI: 10.1056/NEJMoa1602489

[30] Cavalcante P, Le Panse R, Berrih-Aknin S, et al. The thymus in myasthenia gravis: Site of "innate autoimmunity"? Muscle & Nerve. 2011;**44**(4):467-484. DOI: 10.1002/mus.22103

[31] Domeier PP, Schell SL, Rahman ZS. Spontaneous germinal centers and autoimmunity. Autoimmunity. 2017;**50**(1):4-18. DOI: 10.1080/08916934.2017.1280671

[32] Pitzalis C, Jones GW, Bombardieri M, et al. Ectopic lymphoid-like structures in infection, cancer and autoimmunity. Nature Reviews. Immunology. 2014;**14**(7):447-462. DOI: 10.1038/nri3700

[33] Aloisi F, Pujol-Borrell R. Lymphoid neogenesis in chronic inflammatory diseases. Nature Reviews. Immunology. 2006;**6**(3):205-217. DOI: 10.1038/nri1786

[34] Berrih-Aknin S, Ruhlmann N, Bismuth J, et al. CCL21 overexpressed on lymphatic vessels drives thymic hyperplasia in myasthenia. Annals of Neurology. 2009;**66**(4):521-531. DOI: 10.1002/ana.21628

[35] Roxanis I, Micklem K, McConville J, et al. Thymic myoid cells and germinal center formation in myasthenia gravis; possible roles in pathogenesis. Journal of Neuroimmunology. 2002;**125**(1-2):185-197. DOI: 10.1016/s0165-5728(02)00038-3

[36] Schönbeck S, Padberg F, Hohlfeld R, Wekerle H. Transplantation of thymic autoimmune microenvironment to severe combined immunodeficiency mice. A new model of myasthenia gravis. The Journal of Clinical Investigation. 1992;**90**(1):245-250. DOI: 10.1172/JCI115843

[37] Wakkach A, Guyon T, Bruand C, et al. Expression of acetylcholine receptor genes in human thymic epithelial cells: Implications for myasthenia gravis. Journal of Immunology. 1996;**157**(8): 3752-3760

[38] Vincent A, Willcox N, Hill M, et al. Determinant spreading and immune responses to acetylcholine receptors in myasthenia gravis. Immunological Reviews. 1998;**164**:157-168. DOI: 10.1111/j.1600-065x.1998.tb01217.x

[39] Villadangos JA, Heath WR, Carbone FR. Outside looking in: The inner workings of the cross-presentation pathway within dendritic cells. Trends in Immunology. 2007;**28**(2):45-47. DOI: 10.1016/j.it.2006.12.008

[40] Bornemann A, Kirchner T. Thymic myoid cell turnover in myasthenia gravis patients and in normal controls. Virchows Archiv. 1998;**432**(4):357-361. DOI: 10.1007/s004280050178

[41] Leite MI, Jones M, Ströbel P, et al. Myasthenia gravis thymus: Complement vulnerability of epithelial and myoid cells, complement attack on them, and correlations with autoantibody status. The American Journal of Pathology. 2007 Sep;**171**(3):893-905. DOI: 10.2353/ajpath.2007.070240

[42] Poëa-Guyon S, Christadoss P, Le Panse R, et al. Effects of cytokines on acetylcholine receptor expression: Implications for myasthenia gravis. Journal of Immunology. 2005;**174**(10):5941-5949. DOI: 10.4049/jimmunol.174.10.5941

[43] Cron MA, Maillard S, Villegas J, et al. Thymus involvement in early-onset myasthenia gravis. Annals of the New York Academy of Sciences. 2018;**1412**(1):137-145. DOI: 10.1111/nyas.13519

[44] Cufi P, Dragin N, Ruhlmann N, et al. Central role of interferon-beta in thymic events leading to myasthenia gravis. Journal of Autoimmunity. 2014;**52**:44-52. DOI: 10.1016/j.jaut.2013.12.016

[45] Gradolatto A, Nazzal D, Truffault F, et al. Both Treg cells and T conv cells are defective in the myasthenia gravis thymus: Roles of IL-17 and TNF-α. Journal of Autoimmunity. 2014;**52**:53-63. DOI: 10.1016/j.jaut.2013.12.015

[46] Bernard C, Frih H, Pasquet F, et al. Thymoma associated with autoimmune diseases: 85 cases and literature review. Autoimmunity Reviews. 2016;**15**(1):82-92. DOI: 10.1016/j.autrev.2015.09.005

[47] Marx A, Chan JK, Coindre JM, et al. The 2015 World Health Organization classification of tumors of the thymus: Continuity and changes. Journal of Thoracic Oncology. 2015;**10**(10):1383-1395. DOI: 10.1097/JTO.0000000000000654

[48] Maggi L, Andreetta F, Antozzi C, et al. Thymoma-associated myasthenia gravis: outcome, clinical and pathological correlations in 197 patients on a 20-year experience. J Neuroimmunol. 2008;**201-202**:237-244. DOI: 10.1016/j.jneuroim.2008.07.012

[49] Marx A, Willcox N, Leite MI, et al. Thymoma and paraneoplastic myasthenia gravis. Autoimmunity. 2010;**43**(5-6):413-427. DOI: 10.3109/08916930903555935

[50] Savino W, Manganella G, Verley JM, et al. Thymoma epithelial cells secrete thymic hormone but do not express class II antigens of the major histocompatibility complex. The Journal of Clinical Investigation. 1985;**76**(3):1140-1146. DOI: 10.1172/JCI112069

[51] Lefeuvre CM, Payet CA, Fayet OM, et al. Risk factors associated with myasthenia gravis in thymoma patients: The potential role of thymic germinal centers. Journal of Autoimmunity. 2020;**106**:102337. DOI: 10.1016/j.jaut.2019.102337

[52] Wolfe GI, Kaminski HJ, Aban IB, et al. MGTX study group. Long-term effect of thymectomy plus prednisone versus prednisone alone in patients with non-thymomatous myasthenia gravis: 2-year extension of the MGTX randomised trial. Lancet Neurology. 2019;**18**:259-268. DOI: 10.1016/S1474-4422(18)30392-2

[53] Okumura M, Ohta M, Takeuchi Y, et al. The immunologic role of thymectomy in the treatment of myasthenia gravis: Implication of thymus-associated B-lymphocyte subset in reduction of the anti-acetylcholine receptor antibody titer. The Journal of Thoracic and Cardiovascular Surgery. 2003;**126**(6):1922-1928. DOI: 10.1016/s0022-5223(03)00938-3

[54] Fujii Y, Monden Y, Hashimoto J, Nakahara K, Kawashima Y. Acetylcholine receptor antibody-producing cells in thymus and lymph nodes in myasthenia gravis. Clinical Immunology and Immunopathology. 1985;**34**(1):141-146. DOI: 10.1016/0090-1229(85)90018-2

[55] Kawai T, Akira S. The role of pattern-recognition receptors in innate immunity: Update on toll-like receptors. Nature Immunology. 2010;**11**(5):373-384. DOI: 10.1038/ni.1863

[56] Mohammad Hosseini A, Majidi J, Baradaran B, Yousefi M. Toll-like receptors in the pathogenesis of autoimmune diseases. Adv Pharm Bull. 2015;**5**:605-614. DOI: 10.15171/apb.2015.082

[57] Duffy L, O'Reilly SC. Toll-like receptors in the pathogenesis of autoimmune diseases: Recent and emerging translational developments. Immunotargets Ther. 2016;**5**:69-80. Published 2016 Aug 22. DOI: 10.2147/ITT.S89795

[58] Soni C, Wong EB, Domeier PP, et al. B cell-intrinsic TLR7 signaling is essential for the development of spontaneous germinal centers. Journal of Immunology. 2014;**193**(9):4400-4414. DOI: 10.4049/jimmunol.140172

[59] Bernasconi P, Barberis M, Baggi F, et al. Increased toll-like receptor 4 expression in thymus of myasthenic patients with thymitis and thymic involution. The American Journal of Pathology. 2005;**167**(1):129-139. DOI: 10.1016/S0002-9440(10)62960-4

[60] Cordiglieri C, Marolda R, Franzi S, et al. Innate immunity in myasthenia

gravis thymus: Pathogenic effects of toll-like receptor 4 signaling on autoimmunity. Journal of Autoimmunity. 2014;**52**:74-89. DOI: 10.1016/j.jaut.2013.12.013

[61] Cufi P, Dragin N, Weiss JM, et al. Implication of double-stranded RNA signaling in the etiology of autoimmune myasthenia gravis. Annals of Neurology. 2013;**73**(2):281-293. DOI: 10.1002/ana.23791

[62] Robinet M, Maillard S, Cron MA, et al. Review on toll-like receptor activation in myasthenia gravis: Application to the development of new experimental models. Clinical Reviews in Allergy and Immunology. 2017;**52**(1):133-147. DOI: 10.1007/s12016-016-8549-4

[63] Cavalcante P, Galbardi B, Franzi S, et al. Increased expression of toll-like receptors 7 and 9 in myasthenia gravis thymus characterized by active Epstein-Barr virus infection. Immunobiology. 2016;**221**(4):516-527. DOI: 10.1016/j.imbio.2015.12.007

[64] Hennessy EJ, Parker AE, O'Neill LA. Targeting toll-like receptors: Emerging therapeutics? Nature Reviews. Drug Discovery. 2010;**9**(4):293-307. DOI: 10.1038/nrd3203

[65] Thorley-Lawson DA. Epstein-Barr virus: Exploiting the immune system. Nature Reviews. Immunology. 2001;**1**(1):75-82. DOI: 10.1038/35095584

[66] Ning S. Innate immune modulation in EBV infection. Herpesviridae. 2011;**2**(1):1. DOI: 10.1186/2042-4280-2-1

[67] Tattevin P, Le Tulzo Y, Minjolle S, et al. Increasing incidence of severe Epstein-Barr virus-related infectious mononucleosis: Surveillance study. Journal of Clinical Microbiology. 2006;**44**(5):1873-1874. DOI: 10.1128/JCM.44.5.1873-1874.2006

[68] Lassmann H, Niedobitek G, Aloisi F, Middeldorp JM, NeuroproMiSe EBV Working Group. Epstein-Barr virus in the multiple sclerosis brain: A controversial issue-report on a focused workshop held in the Centre for Brain Research of the Medical University of Vienna, Austria. Brain. 2011;**134**: 2772-2786. DOI: 10.1093/brain/awr197

[69] Kalla M, Hammerschmidt W. Human B cells on their route to latent infection-early but transient expression of lytic genes of Epstein-Barr virus. European Journal of Cell Biology. 2012;**91**(1):65-69. DOI: 10.1016/j.ejcb.2011.01.014

[70] Niller HH, Wolf H, Minarovits J. Regulation and dysregulation of Epstein-Barr virus latency: Implications for the development of autoimmune diseases. Autoimmunity. 2008;**41**(4):298-328. DOI: 10.1080/08916930802024772

[71] Henderson S, Huen D, Rowe M, Dawson C, Johnson G, Rickinson A. Epstein-Barr virus-coded BHRF1 protein, a viral homologue of Bcl-2, protects human B cells from programmed cell death. Proceedings of the National Academy of Sciences of the United States of America. 1993;**90**(18):8479-8483. DOI: 10.1073/pnas.90.18.8479

[72] Iizasa H, Kim H, Kartika AV, Kanehiro Y, Yoshiyama H. Role of viral and host microRNAs in immune regulation of Epstein-Barr virus-associated diseases. Frontiers in Immunology. 2020;**11**:367. DOI: 10.3389/fimmu.2020.00367

[73] Taylor GS, Long HM, Brooks JM, et al. The immunology of Epstein-Barr virus-induced disease. Annual Review of Immunology. 2015;**33**:787-821. DOI: 10.1146/annurev-immunol-032414-112326

[74] Serafini B, Rosicarelli B, Franciotta D, et al. Dysregulated Epstein-Barr virus infection in the multiple sclerosis brain.

The Journal of Experimental Medicine. 2007;**204**(12):2899-2912. DOI: 10.1084/jem.20071030

[75] Croia C, Astorri E, Murray-Brown W, et al. Implication of Epstein-Barr virus infection in disease-specific autoreactive B cell activation in ectopic lymphoid structures of Sjögren's syndrome. Arthritis & Rhematology. 2014;**66**(9):2545-2557. DOI: 10.1002/art.38726

[76] Croia C, Serafini B, Bombardieri M, et al. Epstein-Barr virus persistence and infection of autoreactive plasma cells in synovial lymphoid structures in rheumatoid arthritis. Annals of the Rheumatic Diseases. 2013;**72**(9):1559-1568. DOI: 10.1136/annrheumdis-2012-202352

[77] Münz C, Lünemann JD, Getts MT, et al. Antiviral immune responses: Triggers of or triggered by autoimmunity? Nature Reviews. Immunology. 2009;**9**(4):246-258. DOI: 10.1038/nri2527

[78] Dreyfus DH. Genetics and molecular biology of Epstein-Barr virus-encoded BART MicroRNA: A paradigm for viral modulation of host immune response genes and genome stability. Journal of Immunology Research. 2017;**2017**:4758539. DOI: 10.1155/2017/4758539

[79] Lünemann JD, Edwards N, Muraro PA, et al. Increased frequency and broadened specificity of latent EBV nuclear antigen-1-specific T cells in multiple sclerosis. Brain. 2006;**129**:1493-1506. DOI: 10.1093/brain/awl067

[80] Lünemann JD, Jelcić I, Roberts S, et al. EBNA1-specific T cells from patients with multiple sclerosis cross react with myelin antigens and co-produce IFN-gamma and IL-2. The Journal of Experimental Medicine. 2008;**205**(8):1763-1773. DOI: 10.1084/jem.20072397

[81] Poole BD, Gross T, Maier S, Harley JB, James JA. Lupus-like autoantibody development in rabbits and mice after immunization with EBNA-1 fragments. Journal of Autoimmunity. 2008;**31**(4):362-371. DOI: 10.1016/j.jaut.2008.08.007

[82] Balandraud N, Roudier J. Epstein-Barr virus and rheumatoid arthritis. Joint, Bone, Spine. 2018;**85**(2):165-170. DOI: 10.1016/j.jbspin.2017.04.011

[83] Tzartos JS, Khan G, Vossenkamper A, et al. Association of innate immune activation with latent Epstein-Barr virus in active MS lesions. Neurology. 2012;**78**(1):15-23. DOI: 10.1212/WNL.0b013e31823ed057

[84] Severa M, Giacomini E, Gafa V, et al. EBV stimulates TLR- and autophagy-dependent pathways and impairs maturation in plasmacytoid dendritic cells: Implications for viral immune escape. European Journal of Immunology. 2013;**43**(1):147-158. DOI: 10.1002/eji.201242552

[85] Cavalcante P, Serafini B, Rosicarelli B, et al. Epstein-Barr virus persistence and reactivation in myasthenia gravis thymus. Annals of Neurology. 2010;**67**(6):726-738. DOI: 10.1002/ana.21902

[86] Cavalcante P, Maggi L, Colleoni L, et al. Inflammation and Epstein-Barr virus infection are common features of myasthenia gravis thymus: Possible roles in pathogenesis. Autoimmune Dis. 2011;**2011**:213092. DOI: 10.4061/2011/213092

[87] Cavalcante P, Marcuzzo S, Franzi S, et al. Epstein-Barr virus in tumor-infiltrating B cells of myasthenia gravis thymoma: An innocent bystander or an autoimmunity mediator? Oncotarget. 2017;**8**(56):95432-95449. DOI: 10.18632/oncotarget.20731

[88] Klavinskis LS, Willcox N, Oxford JS, et al. Antivirus antibodies in myasthenia

gravis. Neurology. 1985;**35**(9):1381-1384. DOI: 10.1212/wnl.35.9.1381

[89] Bhibhatbhan A, Kline G, Vincent A, et al. Anti-MuSK myasthenia gravis presenting with Epstein-Barr virus-associated mononucleosis and immune-mediated diabetes mellitus. Muscle & Nerve. 2007;**36**(2):264-266. DOI: 10.1002/mus.20746

[90] Csuka D, Banati M, Rozsa C, et al. High anti-EBNA-1 IgG levels are associated with early-onset myasthenia gravis. European Journal of Neurology. 2012;**19**(6):842-846. DOI: 10.1111/j.1468-1331.2011.03636.x

[91] McGuire LJ, Huang DP, Teoh R, et al. Epstein-Barr virus genome in thymoma and thymic lymphoid hyperplasia. The American Journal of Pathology. 1988;**131**(3):385-390

[92] Aoki T, Drachman DB, Asher DM, et al. Attempts to implicate viruses in myasthenia gravis. Neurology. 1985;**35**(2):185-192. DOI: 10.1212/wnl.35.2.185

[93] Klavinskis LS, Willcox HN, Richmond JE, et al. Attempted isolation of viruses from myasthenia gravis thymus. Journal of Neuroimmunology. 1986;**11**(4):287-299. DOI: 10.1016/0165-5728(86)90082-2

[94] Meyer M, Höls AK, Liersch B, et al. Lack of evidence for Epstein-Barr virus infection in myasthenia gravis thymus. Annals of Neurology. 2011;**70**(3):515-518. DOI: 10.1002/ana.22522

[95] Kakalacheva K, Maurer MA, Tackenberg B, et al. Intrathymic Epstein-Barr virus infection is not a prominent feature of myasthenia gravis. Annals of Neurology. 2011;**70**(3):508-514. DOI: 10.1002/ana.22488

[96] Serafini B, Cavalcante P, Bernasconi P, et al. Epstein-Barr virus in myasthenia gravis thymus: A matter of debate. Annals of Neurology. 2011;**70**(3):519. DOI: 10.1002/ana.22544

[97] Iskra S, Kalla M, Delecluse HJ, et al. Toll-like receptor agonists synergistically increase proliferation and activation of B cells by Epstein-Barr virus. Journal of Virology. 2010;**84**(7):3612-3623. DOI: 10.1128/JVI.01400-09

[98] Inghirami G, Chilosi M, Knowles DM. Western thymomas lack Epstein-Barr virus by southern blotting analysis and by polymerase chain reaction. American journal of pathology. 1990;**136**:1429-1436

[99] Engel PJ. Absence of latent Epstein-Barr virus in thymic epithelial tumors as demonstrated by Epstein-Barr-encoded RNA(EBER) in situ hybridization. APMIS. 2000;**108**:393-397. DOI: 10.1034/j.1600-0463.2000.d01-74.x

[100] Chen PC, Pan CC, Yang AH, et al. Detection of Epstein-Barr virus genome within thymic epithelial tumours in Taiwanese patients by nested PCR, PCR in situ hybridization, and RNA in situ hybridization. The Journal of Pathology. 2002;**197**(5):684-688. DOI: 10.1002/path.1141

[101] Takeuchi H, Fujita H, Iwasaki F, et al. A case of Epstein- Barr virus (EBV)-associated thymic carcinoid and investigation of existence of EBV-infected cells in thymus and thymic tumors. Journal of Clinical Microbiology. 2004;**42**(6):2850-2854. DOI: 10.1128/JCM.42.6.2850-2854.2004

[102] Cufi P, Soussan P, Truffault F, et al. Thymoma-associated myasthenia gravis: On the search for a pathogen signature. Journal of Autoimmunity. 2014;**52**:29-35. DOI: 10.1016/j.jaut.2013.12.018

Chapter 6

Epstein-Barr Virus: Should We Still Invest in Vaccines or Focus on Predictive Tests?

Emmanuel Drouet

Abstract

The complex interplay between host and EBV has made it difficult to elaborate useful vaccines protecting against EBV diseases. It is encouraging to see that EBV vaccine programs have started to incorporate different arms of the immune system. An array of argument calls for a realistic goal for vaccine strategies which should be preventing EBV diseases, rather than EBV infection. EBV is the primary cause of infectious mononucleosis and is associated with epithelial cell carcinomas, as well as lymphoid malignancies. Parallel to this need, one could propose priorities for future research: (i) identification of surrogate predictive markers for the development of EBV diseases (ii) determination of immune correlates of protection in animal models and humans.

Keywords: vaccine, early diagnosis, lymphomas, EBV diseases

1. Introduction

More than 95% of the world's adult population is infected with the Epstein–Barr Virus (EBV or HHV4), a virus belonging to the *Herpesviridae* family that mainly infects B lymphocytes. Human Herpesviruses (HHV1–8) have co-evolved through persistent infections in their hosts which are spread efficiently to others and generally do not cause serious disease (**Table 1**) [1]. EBV is transmitted by saliva, usually infects its host during infancy and is largely asymptomatic. If the infection does occur later, in adolescence or young adulthood, in about 40% of cases it leads to the development of an acute condition called infectious mononucleosis (IM). In the United States alone, 125,000 cases of IM are reported each year. EBV is also associated with the development of several malignancies derived either from lymphocytes or from epithelial cells (**Table 1**). It is estimated that about 10% of cancers associated with a viral infection are associated with EBV and that each year, on average, about 200,000 new cases of EBV-associated cancers are diagnosed worldwide [2]. Furthermore, EBV is also associated with the development of autoimmune diseases such as multiple sclerosis [3]. The complex interplay between the Herpesviruses and their hosts has made it difficult to elaborate useful vaccine strategies to protect against HHV-associated diseases [4]. Over the years, the development of HHV vaccines has been a story of mixed fortunes, especially for HSV-2 and HCMV (**Table 2**). The frequent presence of EBV in many pathologies is an indicator of the necessity of developing a vaccine against EBV. The argument was first put forward

Subfamily Herpesviridae	Common abbreviation	Common name	Common manifestations	Antiviral therapy
Alpha-	HSV-1	Herpes Simplex Virus type 1	Cold sores, keratitis, encephalitis	+++
Alpha-	HSV-2	Herpes Simplex Virus type 2	Genital sores	+++
Alpha-	VZV	Varicella Zoster Virus	Chicken pox, shingles	+++
Beta-	HCMV	Human Cytomegalovirus	Severe diseases in the immunocompromised host	++
Beta-	HHV-6	Human Herpesvirus-6	Roseola infantum, rash & fever	—
Beta-	HHV-7	Human Herpesvirus-7	Roseola infantum, rash & fever	—
Gamma-	EBV	Epstein-Barr Virus	Infectious mononucleosis, lymphoid malignancies, nasopharyngeal & gastric carcinoma	+/−
Gamma-	HHV-8	Human Herpesvirus-8	Kaposi sarcoma	—

+++ widely used and successful, ++ widely used and quite successful, + occasionally used with limited success, − rarely used with an uncertain outcome.

Table 1.
List of the major herpesviruses pathogenic for humans. First the alpha- including neurotropic viruses, second the beta- with the most salient virus, CMV. This virus infects a large number of cells and is responsible for a lot of serious diseases in the immunocompromised hosts. HHV6 and HHV7 are lymphotropic viruses, responsible of roseola, and rash and fever in adults. Finally, the gamma- which includes B lymphotropic viruses with transforming activities.

more than 40 years ago by Sir Antony Epstein, the pathologist and expert electron microscopist who discovered the EBV virus [5]. However, to date, despite sustained efforts, the EBV vaccine has not been finalized, even though promising results have been obtained [6, 7]. The main difficulty in developing an anti-EBV vaccine stems from the patchy nature of our knowledge of the course of EBV infection *in vivo.*

Below we review the history of the EBV vaccine development, and current strategies involved. At the same time, one could propose priorities in terms of future research, such as (i) a better definition of the goal for an EBV vaccine; and (ii) the identification of costless surrogate markers that predict the development of EBV-related malignancies.

2. The first challenge for EBV vaccines: the complexity of the biological cycle of EBV

The EBV lifecycle is considerably complex and passes through a phase of latent infection during which the virus induces the activation, proliferation, and differentiation of primary B cells into memory B cells [8, 9]. During this phase, the infection elicits a humoral and cellular immune response directed against the proteins of the latent phase. During the terminal differentiation into plasma cells of infected cells, the productive viral cycle is activated and virions are produced which will be able to infect epithelial cells capable of producing a large number of viral particles. The numerous viral proteins expressed during the production cycle are also important targets of the cellular immune response [10]. The EBV encodes

Herpesvirus	Site of latency and persistence	Pathology	Vaccine trials	Antivirals	Prevalence	Transmission
HHV1 (HSV1)	Neurons (sensory ganglia)	Widespread vesicular lesions and neurological diseases	No ongoing vaccine research	YES	High	Skin contact
HHV2 (HSV2)	Neurons (sensory ganglia)	One of the most prevalent sexually transmitted infections worldwide	In clinical trials no regulatory-approved vaccines	YES	High	Sexual
HHV3 (VZV)	Neurons (sensory ganglia)	Chickenpox	Live vaccine available	YES	High	Respiratory tract
HHV4 (EBV)	B Lymphocytes (oropharyngeal epithelium)	IM, lymphoid & epithelial tumors	In clinical trials no regulatory-approved vaccines	YES	Very high	Saliva
HHV5 (CMV)	Blood monocytes /bone marrow precursors (probably epithelial cells)	Significant disease in pregnancy and immunocompromised patients	In clinical trials no regulatory-approved vaccines	YES	High	Sexual, blood, saliva, urine
HHV6	Monocytes, T lymphocytes	Roseola infantum	—	Not relevant		
HHV7	Monocytes, T lymphocytes	Roseola infantum	—	Not relevant		
HHV8 (KSHV)	Uncertain	Kaposi's sarcoma	No ongoing vaccine research	Not relevant	Moderate	Sexual

Table 2.
Main features of the HHVs. Some vaccines exist, for example the Varilrix© and Zostavax© against VZV. Clinical studies about some other vaccines are in progress (CMV and EBV).

approximately 80 proteins, 15 of which possess at least 90 antigenic epitopes. A large quantity of these proteins stimulates the T-cell receptors (TCRs), but a few interact with the B-cell receptors (BCRs) [11]. Activation of B cells and subsequent antibody production has not only been related to at least 3 envelope glycoproteins (mostly gp350) but also to latency-associated membrane proteins (LMPs). The majority of EBV epitopes inducing either cytotoxic and/or helper T lymphocytes were located on non-structural and/or latency associated proteins. In acute IM patients (approximately 40%), a considerable proportion of HLA B8 restricted CTL reactivity is directed against a single peptide (RAKFKQLL) of the *trans*-activator protein BZLF1/Zta/ZEBRA [10].

It must be noted that natural killer cells and anti-EBV cytotoxic T lymphocytes (CTLs) are the main players in the immune response that effectively controls infection [12]. The primary role of anti-EBV CTLs would be to control the proliferation

Vaccine candidates	EBV antigens used	Results
Epitope vaccine	EBNA3A	Induced EBNA3A-specific T-cell responses Did not protect against EBV infection. Lower incidence of IM in vaccinated people
Antigen–antibody conjugates	EBNA1 Several latent antigens	Targeting of DC enabled the induction of EBNA-1spcefic CD4+ and CD8+ T cells Vaccination of humanized mice generated EBNA1 specific T cells
Monomeric	gp350	Induced neutralizing gp350-specific antibodies. Reduced incidence of IM Did not protect against EBV infection.
Multimeric	Tetrameric gp350 Trimeric gH/gL and trimeric gB	Higher immunity (neutralizing antibodies) of tetrameric gp350 compared to that of monomeric one (vaccinated rabbits). Higher immunity (neutralizing antibodies) of trimeric gH/gL and trimeric gB compared to that of monomeric gp350 (vaccinated rabbits).
Nanoparticles	gp350	Higher immunity (neutralizing antibodies) of gp350-containing nanoparticles, compared to that of monomeric gp350 (vaccinated mice and monkeys).
Chimeric NDV VLPs	gp350 gH/gL, gp42, LMP2 & EBNA1	Higher immunity (neutralizing antibodies) of gp350-containing NDV*-VLPs, compared to that of monomeric gp350 (vaccinated mice). Use of NDV*-VLP platform to combine EBV lytic and latent antigens
EBV VLPs	More than three dozen structural proteins More than three dozen structural proteins & EBNA1	Similar antigenicity with wt EBV Increased protection against EBV infection (humanized mice)
Recombinant adenovirus	ZEBRA immediate-early protein (BZLF1 gene)	Prolonged survival from fatal EBV-LPD (humanized mice)
mRNA	mRNA-1189 (gp350-gH/gL-gB) Moderna Inc. platform	Antibody titers against viral proteins involved in epithelial cell entry (gH/gLand gB) or B cell entry (gp350, gH/gLand gB) were measured in peripheral blood at day 57 (mice)

Table 3.
*Summary of prophylactic EBV vaccine candidates that have been developed (from ref. [7]) * NDV Newcastle disease.*

of latently infected B cells. EBV has the feature of implementing various latency and lytic transcription programs, suggesting that it assumes distinct antigenic states within infected individuals (**Table 3**) [7, 10]. Yet, despite the wide variety of antigens that predominate throughout the EBV life cycle, EBV vaccine candidates have traditionally only focused on a limited number of EBV antigens. For a summary on these vaccine candidates see a review by Cohen [13]. Thus, the EBV vaccines designed so far fall into two categories: those preventing any kind of infection (including prophylaxis of EBV-associated malignancies) and those designed for therapeutic purposes (to be used in subjects already infected) [13].

3. A second challenge: the lack of a true animal model

EBV is highly species-specific and only infects humans and primates. Initial studies of EBV vaccines used cottontop tamarins (white-crested), a now endangered species [14]. This model has several drawbacks including the very high doses of EBV in the challenge inoculum required to cause tumors, a non-physiologic route of infection (intraperitoneal injection of virus), and the fact that EBV is not a natural pathogen in these animals. Moreover, EBV does not establish a latent infection in the B cells of these monkeys which is the case with humans. EBV vaccine studies have also been performed in common marmosets and EBV gp350 can protect against the parenteral challenge of these animals [15]. The use of an animal virus such as *Rhesus* lymphocryptovirus (LCV) is also a useful model for vaccination studies [16] as it is close to EBV and reproduces the majority of EBV symptoms in its natural host (the *Rhesus* macaque). Today, the humanized mouse model looks promising, as it is possible to recreate different pathologies associated with EBV [17–19]. However, the absence of infection in the epithelial cells of the animal does not allow the reconstitution of all the pathologies associated with EBV in humans. Two types of vaccines are currently being studied to control EBV: A prophylactic vaccine that aims to neutralize the virus to block infection and a therapeutic vaccine that aims to induce or improve the anti-EBV cellular response in some patients.

4. The prophylactic vaccines

In designing a prophylactic vaccine against a virus, induction of a neutralizing antibody response is generally sought. Multiple alternative vaccine candidates include targeting EBV-based glycoproteins, EBV lytic proteins, and EBV latent proteins (**Figure 1**).

4.1 Which antigens are used?

The glycoproteins of the viral envelope are therefore the preferred targets. EBV carries several glycoproteins (gp350, gB, gH, gL, gp42, gM, gN, BMRF2, BDLF2, BDLF3, BILF2, BILF1, BARF1) [20] on its surface. To date, the greatest strides towards developing an EBV-based vaccine have been made by targeting gp350. This type I glycoprotein is crucial for EBV's ability to enter the host B cells by binding their CD21 or CD35 receptor. It is the most abundant glycoprotein on the surface of virions and the most expressed by cells infected with EBV. In addition, gp350 is the major target of antibodies capable of neutralizing infection of B cells [20]; it is also an important antigen and target recognized by cellular immunity [10]. Other EBV glycoproteins such as (i) gH/gL (member of the fusion complex); (ii) gp42 (determines the cellular tropism of EBV); and (iii) gB (a class III type of viral fusion protein

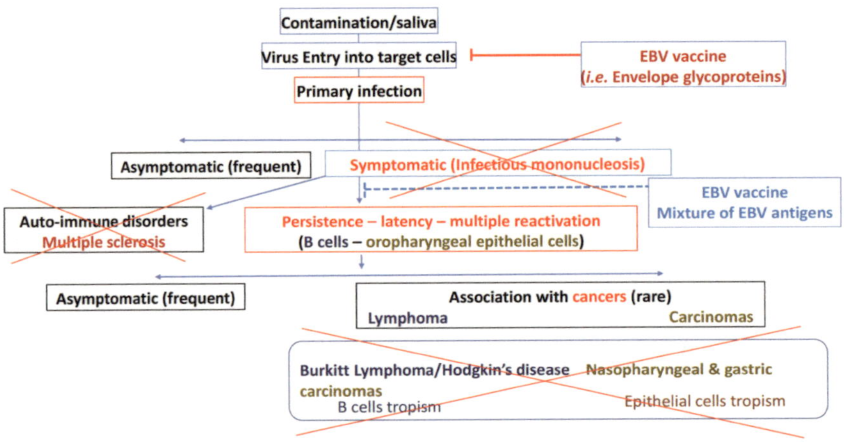

Figure 1.
An EBV vaccination strategy: Whatever EBV vaccine candidates, they will not block infection. However, they could prevent the onset of the symptoms of infection and reduce the risk of developing EBV-associated tumors or EBV-associated pathology (i.e. MS).

also essential for maturation and egress) have also been tested in various vaccine trials [7]. They are also able to induce the expression of neutralizing antibodies. They do, however, appear to be less effective than those directed against gp350 (see **Table 3**).

For the various reasons cited above, most vaccination trials have been carried out using recombinant gp350 [6, 7, 13]. The production of viral-like particles (VLPs) which do not contain a genome and in which several viral proteins involved in cell transformation have been inactivated or deleted has also been described. The vaccination of mice with these VLPs induces the production of neutralizing antibodies and cellular immune response [7, 21]. Although, while this approach is promising, given the oncogenic properties of EBV, the acceptability of this type of vaccine for use in human health remains uncertain.

4.2 The clinical trials

Several vaccines have been evaluated to prevent infection and protect against the symptomatic episode of primary infection (IM) [13]. The first vaccine trial in humans was undertaken by Gu and colleagues [22] using a live recombinant vaccinia virus/major EBV envelope antigen BNLF-1 MA (gp220–340) construct. The authors showed that for the first time it was possible to protect against and/or delay EBV infection by the natural route. The most advanced study of the safety, immunogenicity and efficacy of an EBV vaccine has been reported by Sokal and colleagues [23]. In their study (NCT00430534), a group of 181 EBV seronegative volunteers between the ages of 16 and 25 received three doses of a recombinant gp350 vaccine or placebo. The authors claimed the vaccine had demonstrable efficacy (mean efficacy rate, 78.0% [95% confidence interval: 1.0–96%]) and that there were no concerns regarding safety or immunogenicity. Over an observation period of 18 months, the vaccination of the young adults with recombinant gp350 reduced the risk of developing an IM of 10% in controls to only 2% in vaccinated people. However, despite the production of neutralizing antibodies, vaccination does not appear to prevent infection. This result suggests that the vaccine used may reduce the risk of associated pathogenesis without necessarily preventing infection. The same type of vaccine that was given to patients not infected with EBV, and who were waiting for a kidney transplant, did not seem to give satisfactory results (only 30% produced neutralizing antibodies). The unsatisfactory results were probably because patients in this study suffered from both an immunosuppressed state and renal impairment [24].

Alternative EBV vaccines such as, a virus-like particle EBV vaccine [21] and a CD8+ T-cell peptide epitope-based vaccine [25]. have been evaluated in Phase 1 clinical trials. To date, several phases I and phase II clinical trials have been carried out and have yielded rather encouraging results [7, 26].

In 10% of cases, IM is associated with quite severe symptoms (such as prolonged asthenia, risk of Hodgkin's lymphoma) [27–29] and in the most serious cases with neurological involvement (1%) [3, 30]. Given the vast variability in these results and the long period from an EBV infection to the onset of MS, such EBV vaccine trials in MS populations are not feasible. With a 30 times greater rate of MS occurrence in first-degree relatives when compared to unrelated populations, such intervention may potentially decrease the overall MS incidence [31]. This explains the increasing interest in developing EBV vaccines to prevent MS. Given the association of IM with MS, there is a strong possibility of reducing this childhood infection by eradicating MS. [32]. So far, however, there is no licensed EBV vaccine and to make progress regarding its development, a greater understanding of the association of EBV with MS is required [33]. Recent advances have pointed to the use of EBV vaccines to control MS. Indeed, both asymptomatic EBV infection and IM have also been associated with an increase in MS susceptibility [33, 34]. The imminent increase in MS risk following an EBV seroconversion has been expertly shown through a study that utilized serial blood samples derived from more than 8 million active-duty military personnel [35].

Other potential targets for vaccine development include immediate and early EBV proteins that are expressed during the first steps of the lytic cycle. Both Zta and Rta immediate proteins (encoded by BZLF1 and BRLF1, respectively) are easily recognizable due to an uninhibited CD8+ T-cell response [36]. On the other hand, the early lytic proteins BMLF1 and BMRF1 can be detected by CD4+ T cells as early as the first day of EBV infection [37]. Studies have examined the utility of the BZLF1 vaccine in mice models of EBV-induced post-transplant lymphoproliferative disorder and shown successful T-cell immunity induction towards the infected tumor cells [38]. Lastly, recent evidence also shows that the latent proteins (EBNA) can be recognized by CD8+ and CD4+ T cells and prevent further expansion of EBV-infected B cells [37]. As we now understand the importance of B cells in the MS pathophysiology, we can conclude that this type of vaccine intervention would potentially exert a therapeutic outcome [39]. In contrast, an effective EBV vaccine that could prevent the 200,000 new EBV-associated malignancies occurring globally each year is not currently available despite the considerable efforts expended in developing EBV gp350 vaccines [6]. Very recently, in 2020, the Moderna Company (Cambridge, MA, USA) carried an innovative mRNA-based EBV vaccine (mRNA-1189) (https://investors.modernatx.com/static-files/834b6509-553f-4ee5-84e0-c198bbb850f0). Preclinical data demonstrated the ability to induce antibodies against EBV antigens: Naïve Balb/c mice were given two doses of a vaccine against EBV antigens in combination approximately 4 weeks apart. Antibody titers against viral proteins involved in epithelial cell entry (gH/gL and gB) or B cell entry (gp350, gH/gL and gB) (**Table 3**) were measured in peripheral blood at day 57. Their last results demonstrated high levels of anti-EBV neutralizing antibodies, and at levels significantly higher than those observed in naturally-infected human sera.

4.3 Major drawbacks of the prophylactic EBV vaccine strategies

Contrary to Epstein's initial idea, an EBV vaccination does not block infection. However, it could prevent the onset of the symptoms of infection and reduce the risk of developing EBV-associated tumors. Moreover, the correlates of protection against EBV infection and diseases (in animal models and humans) are not clearly defined, so it is hard to reliably predict the ideal EBV vaccine targets and whether humoral immunity or cellular immunity or both should be involved. Currently, the definitions of a goal for an EBV vaccine and criteria for licensure to prevent diseases rather than infections are not clear.

5. The therapeutic vaccines

In the case of a therapeutic vaccine, the induction of cellular immunity against EBV in patients suffering from certain pathologies (NPC, HD, etc.) is the main objective. One of the difficulties of this approach is that the number of viral proteins expressed in cancer cells varies according to the pathology concerned. The EBNA-1 protein is the only viral protein expressed in all cases of EBV-associated cancers. It is also one of the main targets of CD4 T cells along with the membrane proteins LMP1 and LMP2 which are also good targets for CD8 T cells. This makes this type of approach possible. The relevance of designing a therapeutic anti-EBV vaccine is based on clinical observations from tests of infusions of EBV-specific T lymphocytes (CTLs directed against the viral proteins LMP1 and LMP2). In patients with Hodgkin's lymphoma, non-Hodgkin's lymphoma or nasopharyngeal carcinoma (NPC), the results of the first studies are encouraging - this specific EBV cell therapy can markedly improve the survival of some of these patients [7, 40–42]. Therefore, a vaccine that induces T-cell responses to tumor-expressed EBV latency proteins could improve patient survival.

In the context of cure therapy, the adoptive transfer of EBV-specific T cells has been therapeutically explored for decades with clinical success [43]. To avoid naturally occurring EBV-specific autologous T-cell selection from every patient, the transgenic expression of latent and early lytic viral antigen-specific TCRs essential to redirect T cells and to target the respective tumors has been investigated. The latest evidence suggests that not only TCRs against transforming latent EBV antigens, but also against early lytic viral gene products, might be protective for the control of EBV infection and associated oncogenesis [44]. At the same time, these approaches might be more selective and cause less collateral damage rather than targeting general B-cell markers with chimeric antigen receptors (CARs). Thus, EBV-specific TCR transgenic T cells constitute a brilliant therapeutic strategy against EBV-associated malignancies [45]. As an example, recent studies describing CD8^{+}gp350CAR-T cells showed proof-of-concept preclinical efficacy against impending EBV^{+} lymphoproliferation and lymphomagenesis [46].

6. Uncertainties surrounding EBV vaccines

Despite the very encouraging results obtained in phase II clinical trials, to date, no phase III trials have been implemented. The reason why no further development of this vaccine has been done is not known. Given the large diversity of pathologies associated with EBV, it is unlikely that a single vaccine applicable to all diseases associated with EBV can be developed. Vaccination against EBV must take into account various factors such as the geographic characteristics of certain pathologies (NPC in South-East Asia, endemic Burkitt's lymphoma (BL) in equatorial Africa), the incubation period necessary between infection and development of the disease pathology (IM: 4 to 6 weeks, NPC: > 30 years), and the initial age of infection. Such disparity complicates vaccination strategies which would need to be implemented. Despite this and depending on the pathology involved, it is still worth considering further research on an anti-EBV vaccination program. According to a recent US study among university students, 37% were EBV negative when they entered university, but 3 years later 46% of them had experienced EBV seroconversion. Of these, 77% went on to develop an NID. It is interesting to note that IM is the most common cause of absenteeism among new recruits to the US military. In developed countries, these epidemiological observations support the idea of administering a vaccine capable of preventing the disease when administered to children aged between 11 and 12 who are EBV negative (in tandem with the administration of a vaccine such as a papillomavirus). Such a vaccine would reduce the risk of developing IM and would also reduce the risk of developing Hodgkin lymphoma [47] or MS [30], pathologies which are linked to EBV and which are most likely to be a consequence of the EBV-induced immunological disorder in IM [48].

The value of an EBV vaccination to protect children against BL, especially in Africa, is certainly significant. Nevertheless, in this region of the world, infection with EBV generally occurs early (50% of children are infected by their 1st year), and it is certainly not easy to vaccinate at this stage, especially if three injections are required to achieve protective immunity. It is, however, not impossible and has been implemented in some countries where children are given vaccinations which include hepatitis B in early childhood.

EBV is associated with various lympho-proliferations in immunocompromised people, especially following transplantation, or HIV infection. The risk of developing PTLD is 25 to 30 times higher in an EBV-negative person before the transplant than in a person who was HIV-positive. Prophylactic vaccination against EBV would not only reduce the risks associated with the primary infection but could also decrease the risk of developing PTLD during transplantation; the latter hypothesis has not yet been evaluated. Regarding NPC and gastric carcinoma, only retrospective studies after prophylactic vaccination could reveal its efficacy. It would then be necessary to demonstrate the direct effects of a prophylactic vaccine aimed at preventing these pathologies which develop more than 30 years after infection. Nevertheless, this has already been achieved with the hepatitis B vaccination program which is performed in children and protects against the development of liver cancer 15 to 20 years later.

7. The EBV diagnostic tests as a predictor of diseases

Taking into account the above-mentioned points, it is undoubtedly time to turn to predictive tests to prevent the appearance of the first signs of pathology both in the context of cancers (lymphomas) and in the context of chronic pathologies.

7.1 What about conventional EBV diagnostic tests?

EBV serology was for a long time the only technique used for diagnosis. In immunocompromised patients, serological tests (looking for IgG and IgM antibodies directed against two types of viral antigens - capsid and EBNAs) are used to identify the immune status to EBV in the donor/recipient (before transplantation) and in HIV patients. They are not used for primary or hereditary immunodeficiencies. With regards to EBV serology, the practice is relatively homogeneous with an assay combining anti-VCA IgG, anti-VCA IgM and anti-EBNA IgG. The combined use of these three markers is sufficient for most diagnoses, making it possible to distinguish the primary infection (±, +, –) from an old infection (+, –, +). Depending on the manufacturers of diagnostic kits, the detection technique (ELISA or immunofluorescence) and the nature of the antigen targets (recombinant proteins, infected cells, peptides, etc.) of these antibodies may vary, but most of the techniques used are validated for their diagnostic capacities by expert medical virology laboratories at the time of marketing.

The ability to accurately determine viral load (DNA PCR) for HHV infections post-transplantation [4] has become a mainstay for diagnostics especially in the context of the beta and gamma herpesviruses. Most approaches use real-time quantitative PCR-based assays [49]. PTLD (classified into six categories by WHO) has become a deleterious complication of both solid-organ and hematopoietic stem-cell transplantation. Data from large transplantation registries have shown an increased incidence of PTLD and significant associations with morbidity and mortality [50, 51]. EBV viral load monitoring is now routine and high viral loads are often associated with concurrent PTLD. But data linking EBV kinetics to the risk of developing PTLD remain controversial. Measurement of EBV viral load by quantitative PCR is an essential test in the follow-up of children with solid organ transplants. It is used: (i) to prevent the development of PTLD (by adapting immunosuppression and/or by initiating pre-emptive treatment); (ii) in the monitoring of pre-emptive treatment; and (iii) in the follow-up of the curative treatment of PTLD.

It should be emphasized that there is a difference between patients who have had solid organ allografting and HSCT patients [50]. Immunosuppressive treatments in solid organ allograft recipients are modest compared to HSCT recipients who receive more severe immunosuppressive treatment. Correlations between higher EBV loads and the development of PTLD are seen in solid organ allograft recipients, but these correlations do not indicate high positive and negative predictive values [52]. There is considerable overlap between the EBV loads in patients with PTLD and those in patients without PTLD. Furthermore, solid organ allograft recipients receive lifelong immunosuppression so that there is a long-term risk of EBV-PTLD. Therefore, routine surveillance for EBV-DNA by quantitative PCR is not recommended in adult recipients [53]. Solid-organ allograft recipients also sometimes carry chronic high EBV loads without symptoms consistent with PTLD [53, 54]. However, the significance of a high EBV load in terms of long-term health is unknown. Conversely, in children at high risk of primary EBV infection, routine surveillance is beneficial for the preemptive identification of patients at high risk of PTLD [53]. Finally, a current article investigating both the EBV DNA load in whole blood and EBV serology in HIV-infected patients with classical Hodgkin concluded that EBV DNA loads at diagnosis were not prognostic [55].

Not unlike the situation with CMV, the lack of an international genome standard for quantification of EBV in molecular assays makes a comparison of thresholds for impending PTLD difficult to interpret [4]. In contrast, EBV infection and in particular EBV-driven PTLD is a more difficult disease to manage with little evidence that antiherpes drugs are effective especially once PTLD is manifest [4]. Anti-CD20 antibodies (Mabthera®) are not introduced until PTLD has been confirmed. In contrast, for other types of transplants (intestines, lung, heart, kidneys), anti-CD20 antibodies can be used earlier and they are part of the pre-emptive treatment. There is currently no consensus on the best preemptive strategy because the threshold, or kinetics, of preemptive intervention is difficult to define. Typically, a one-log increase in viral load is a warning sign. It should be noted that the therapeutic decision is based on a set of arguments which are virological, clinical (such as tonsillar hypertrophy for example) or biological (for example LDH, uric acid)), because the EBV viral load is not the only predictive marker of PTLD.

7.2 New biomarkers and therapies exploring the lytic cycle

For decades, many articles have reported the presence of an EBV lytic cycle in tumor cells from HL, NPC, in transplant patients, and breast tumors. Clinical studies on EBV lytic proteins including ZEBRA in patients with PTLD or HIV-associated non-Hodgkin lymphoma NHL are mostly related to the role of these proteins in neoplastic tissues. Both high EBV copy number and strong BZLF1 mRNA expression in the peripheral blood lymphocytes (PBL) of patients are sensitive markers of EBV-related PTLD. Soluble ZEBRA concentrations of >100 ng/mL detected by an enzyme-linked immunosorbent assay (ELISA) in serum of patients after solid organ or hematopoietic stem cell transplant were predictive of PTLD in 80% of the cases within 3 weeks [56] (for a review see ref. 57). Thus, ZEBRA testing in serum could help identify patients likely to develop severe outcomes during the critical post-transplant period and serve as a potential diagnostic/prognostic marker for EBV follow-up in immunocompromised patients (**Figure 2**).

The relevance of the EBV lytic cycle to human pathology prompted researchers to target certain lytic proteins with therapeutic aims [57]. As an example, adenovirus vectors expressing BZLF1 or BRLF1 were used to treat EBV-positive tumors [58]. On the other hand, the Food and Drug Administration (FDA)-approved leflunomide, which targets EBV replication, was shown to inhibit the earliest step of lytic EBV reactivation (BZLF1 and BMRF1 expression) and prevented the development of EBV-induced lymphomas in both a humanized mouse model and a xenograft model [59]. More recently, duvelisib (a molecule inhibiting the PI3K/AKT signaling pathway, and BCR signaling) was shown to reduce cell growth and expression of EBV lytic genes BZLF1 and gp350/220 in EBV-positive cell lines [60]. The histone deacetylase (HDAC) and DNA methyltransferase inhibitors are also possible avenues to suppress the ZEBRA expression and the entire lytic cascade [61]. To summarize, efforts should be made to improve the relevance of using ZEBRA protein in future EBV vaccine settings [62].

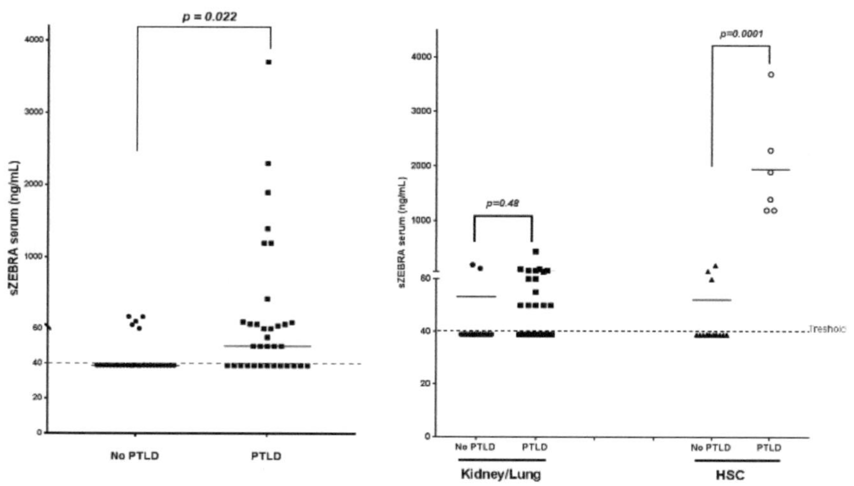

Figure 2.
ZEBRA as a specific marker measuring early activation of replication of the oncogenic EBV, providing more precise monitoring of posttransplant lymphoproliferative disorder development in transplant patients. (from Habib et al. [56] with permission) (HSC = hematopoietic stem cells).

8. Conclusion

The issue of an EBV vaccine is a very actual topics since there is more and more evidence for an association between EBV primary infection (IM) and the development of multiple sclerosis (MS) and Hodgkin's lymphoma. Numerous prophylactic IM vaccines targeting EBV proteins have been developed. They have shown partial success in reducing IM but have failed to prevent EBV infection. Therapeutic vaccines against NPCs have had considerable success, but there is a need to improve their effectiveness. Increasing vaccine activity in NPC (or gastric carcinoma) might be difficult due to a long latency period between primary infection and the development of these carcinomas (see **Table 4**) [63]. The addition of new targets and the recent advances in mRNA vaccines may further improve the efficacy of therapeutic and prophylactic vaccines against EBV [64].

However, the design of a prophylactic vaccine against EBV poses serious problems: It is still difficult to find exact correlates of protection and it is still problematic to define the populations intended to receive the vaccine. Immunotherapeutic strategies, including CAR T cells, are emerging as new platforms for the treatment of tumors associated with EBV [45, 46]. The incorporation of immunotherapeutic strategies as first-line treatment may provide better long-term results. It remains to be seen how the various immunotherapeutic strategies will be incorporated into future therapeutic strategies.

On the other hand, the design of new predictive tests (*i.e.* ZEBRA-based) capable of monitoring the intensity of EBV reactivation and tumor progression, could more easily help the physician to monitor the course of pathologies linked to

Prospects	Progress	Problems
Prevention of infectious mononucleosis	IM was prevented in a phase 2 study with a subunit gp350 vaccine [23] A CD8+ T-cell peptide (EBNA3-TT) vaccine was immunogenic with a hint of efficacy [25]	Duration of protection unknown. Viral loads and T-cell-specific responses were not evaluated. The ideal age' which to vaccinate may differ according to race/ethnicity and socioeconomics
Prevention of Nasopharyngeal Carcinoma	Vaccinia constructs expressing EBV glycoprotein (gp 220–340) are immunogenic and may have reduced incidence of EBV infection in Chinese children	CD8+ T-cell peptide vaccine: HLA restricted. The long incubation period from EBV infection to the development of NPC makes efficacy trials impractical.
Prevention of lymphomas	Subunit gp 350 vaccines are safe in pediatric renal transplant candidates	The vaccine was poorly immunogenic probably due to the low dose and weak adjuvant; the trial could not assess protection from PTLD
Treatment of NPC	Vaccinia recombinant vectors expressing the tumor-associated latent or lytic viral antigens aresafe and immunogenic [41, 42, 58]	Therapeutic efficacy has not yet been assessed
Prevention of multiple sclerosis	Evidence that a vaccine could work: EBV-specific CD8+ T cell responses are elevated during active MS [39].	The long incubation period from EBV infection to MS makes vaccine efficacy trials impractical except perhaps in first-degree relatives

Table 4.
Prospects, progress and problems in EBV vaccine development (from Balfour HH [63]).

DOI: http://dx.doi.org/10.5772/intechopen.101094

EBV replication (*i.e.* lymphomas, MS). Such approaches will easily make it possible to initiate pre-emptive antiviral treatments; in addition, these diagnostic tests have the advantage of being minimally invasive and inexpensive.

Author details

Emmanuel Drouet
Université Grenoble-Alpes, France

*Address all correspondence to: emmanuel.drouet@ibs.fr

References

[1] Roizman B. The family herpesviridae: General description, taxonomy, and classification. In: Roizman B, editor. The Herpesviruses. The Viruses. Boston, MA: Springer; 1982. DOI: 10.1007/978-1-4684-4163-5_1

[2] Farrell PJ. Epstein–Barr virus and cancer. Annual Review of Pathology. 2019;**14**:29-53. DOI: 10.1146/annurev-pathmechdis-012418-013023

[3] Ascherio A, Kassandra L, Munger KL. Epidemiology of multiple sclerosis: From risk factors to prevention—an update. Seminars in Neurology. 2016;**36**(02):103-114. DOI: 10.1055/s-0036-1579693

[4] Emery VC. Human herpesvirus vaccines and future directions. American Journal of Transplantation. 2013;**13**(Suppl 3):79-86; quiz 86. DOI: 10.1111/ajt.1. PMID: 23347216

[5] Epstein MA, Achong BG. The Epstein-Barr virus. Annual Review of Microbiology. 1973;**27**:413-436

[6] Cohen JI. Epstein-Barr virus vaccines. Clinical & Translational Immunology. 2015;**4**(1):e32. DOI: 10.1038/cti.2014.27

[7] Van Zyl DG, Mautner J, Delecluse H-J. Progress in EBV vaccines. Frontiers in Oncology. 2019;**9**:104. DOI: 10.3389/fonc.2019.00104

[8] Longnecker L, Kieff E, Cohen JI. Epstein-Barr Virus. Philadelphia, PA: Lippincott Williams andWilkins; 2013

[9] Price AM, Luftig MA. Dynamic Epstein-Barr virus gene expression on the path to B-cell transformation. Advances in Virus Research. 2014;**88**:279-313. DOI: 10.1016/B978-0-12-800098-4.00006-4

[10] Hislop AD, Taylor GS, Sauce D, Rickinson AB. Cellular responses to viral infection in humans: Lessons from Epstein-Barr virus. Annual Review of Immunology. 2007;**25**:587-617. DOI: 10.1146/annurev.immunol.25.022106.141553 PMID: 17378764

[11] Rajcáni J, Szenthe K, Banati F, Szathmary S. Survey of Epstein Barr Virus (EBV) immunogenic proteins and their epitopes: Implications for vaccine preparation recent pat anti-infect. Drug Discovery. 2014;**9**(1):62-76

[12] Münz C. Epstein Barr virus - a tumor virus that needs cytotoxic lymphocytes to persist asymptomatically. Current Opinion in Virology. 2016;**20**:34-39

[13] Cohen JI. Vaccine development for Epstein-Barr virus. Advances in Experimental Medicine and Biology. 2018;**1045**:477-493

[14] Epstein MA, Morgan AJ, Finerty S, Randle BJ, Kirkwood JK. Protection of cottontop tamarins against Epstein-Barr virus-induced malignant lymphoma by a prototype subunit vaccine. Nature. 1985;**318**:287-289

[15] Emini EA, Schleif WA, Silberklang M, et al. Vero cell-expressed Epstein- Barrvirus (EBV) gp350/220 protects marmosets from EBV challenge. Journal of Medical Virology. 1989;**27**(2):120-123. [PubMed: 2537882]

[16] Morgan AJ, Khanna R. Epstein–Barr virus vaccines. In: Arvin A, Campadelli-Fiume G, Mocarski E, et al., editors. Human Herpesviruses: Biology, Therapy, and Immunoprophylaxis. Cambridge: Cambridge University Press; 2007 Chapter 72

[17] Münz C. EBV infection of mice with reconstituted human immune system components. Current Topics in Microbiology and Immunology. 2015;**391**:407-423. DOI: 10.1007/978-3-319-22834-1_14 PMID: 26428383

[18] Kanekiyo M, Bu W, Joyce MG, et al. Rational design of an Epstein-Barr virus vaccine targeting the receptor-binding site. Cell. 2015;**16002**(5):1090-1100. DOI: 10.1016/j.cell.2015.07.043. Epub 2015 Aug 13. PMID: 26279189; PMCID: PMC4757492

[19] Bu W, Joyce MG, Nguyen H, et al. Immunization with components of the viral fusion apparatus elicits antibodies that neutralize Epstein-Barr virus in B Cells and epithelial cells. Immunity. 2019;**50**(5):1305-1316.e6. DOI: 10.1016/j.immuni.2019.03.010. Epub 2019 Apr 9. PMID: 30979688; PMCID: PMC66 60903

[20] Möhl BS, Chen J, Longnecker R. Gammaherpesvirus entry and fusion: A tale how two human pathogenic viruses enter their host cells. Advances in Virus Research. 2019;**104**:313-343. DOI: 10.1016/bs.aivir.2019.05.006. Epub 2019 Jul 17. PMID: 31439152

[21] Ruiss R, Jochum S, Wanner G, et al. A virus-like particle-based Epstein-Barr virus vaccine. Journal of Virology. 2011;**85**:13105-13113

[22] Gu SY, Huang TM, Ruan L, et al. First EBV vaccine trial in humans using recombinant vaccinia virus expressing the major membrane antigen. Developments in Biological Standardization. 1995;**84**:171-177

[23] Sokal EM, Hoppenbrouwers K, Vandermeulen C, et al. Recombinant gp350 vaccine for infectious mononucleosis: A Phase 2 randomized, double-blind, placebo-controlled trial to evaluate the safety, immunogenicity, and efficacy of an Epstein-Barr virus vaccine in healthy young adults. The Journal of Infectious Diseases. 2007;**196**: 1749-1753

[24] Rees L, Tizard EJ, Morgan AJ, et al. A phase I trial of Epstein-Barr Virus gp350 vaccine for children with chronic kidney disease awaiting transplantation. Transplantation. 2009;**88**:1025-1029

[25] Elliott SL, Suhrbier A, Miles JJ, et al. Phase 1 trial of a CD8+ T-cell peptide epitope-based vaccine for infectious mononucleosis. Journal of Virology. 2008;**82**:1448-1457

[26] Moutschen M, Leonard P, Sokal EM, et al. Phase I/II studies to evaluate safety and immunogenicity of a recombinant gp350 Epstein-Barr virus vaccine in healthy adults. Vaccine. 2007;**25**: 4697-4705

[27] Balfour HH Jr, Dunmire SK, Hogquist KA. Infectious mononucleosis. Clinical & Translational Immunology. 2015;**4**:e33. DOI: 10.1038/cti.2015.1

[28] Buchwald DS, Rea TD, Katon WJ, et al. Acute infectious mononucleosis: Characteristics of patients who report failure to recover. The American Journal of Medicine. 2000;**109**:531-537. DOI: 10.1016/S0002-9343(00)00560-X

[29] Hjalgrim H, Smedby KE, Rostgaard K, et al. Infectious mononucleosis, childhood social environment, and risk of Hodgkin lymphoma. Cancer Research. 2007;**67**:2382-2388. DOI: 10.1158/0008-5472.CAN-06-3566

[30] Handel AE, Williamson AJ, Disanto G, et al. An updated meta-analysis of risk of multiple sclerosis following infectious mononucleosis. PLoS One. 2010;**5**:e12496. DOI: 10.1371/journal.pone.0012496

[31] Xia Z, White CC, Owen EK, et al. Genes and environment in multiple sclerosis project: A platform to investigate multiple sclerosis risk. Annals of Neurology. 2016;**79**:178-189

[32] Balfour HH Jr, Schmeling DO, Grimm-Geris JM. The promise of a prophylactic Epstein-Barr virus vaccine. Pediatric Research. 2019;**87**(2):345-352.

DOI: 10.1038/s41390-019-0591-5 Epub 2019 Oct 3. PMID: 31641280

[33] Guan Y, Jakimovski D, Ramanathan M, et al. The role of Epstein-Barr virus in multiple sclerosis: From molecular pathophysiology to in vivo imaging. Neural Regeneration Research. 2019;**14**:373-386

[34] Ascherio A, Munger KL, Lennette ET, et al. Epstein-Barr virus antibodies and risk of multiple sclerosis: A prospective study. JAMA. 2001; **286**:3083-3088

[35] Levin LI, Munger KL, O'Reilly, et al. Primary infection with the Epstein-Barr virus and risk of multiple sclerosis. Annals of Neurology. 2010;**67**:824-830

[36] Pudney VA, Leese AM, Rickinson AB, Hislop AD. CD8+ immunodominance among Epstein-Barr virus lytic cycle antigens directly reflects the eciency of antigen presentation in lytically infected cells. The Journal of Experimental Medicine. 2005;**201**:349-360

[37] Brooks JM, Long HM, Tierney RJ, et al. Early T Cell recognition of B Cells following Epstein-Barr virus infection: Identifying potential targets for prophylactic vaccination. PLoS Pathogens. 2016;**12**:e1005549

[38] Hartlage AS, Liu T, Patton JT, et al. The Epstein-Barr virus lytic protein BZLF1 as a candidate target antigen for vaccine development. Cancer Immunology Research. 2015;**3**:787-794

[39] Jakimovski D, Weinstock-Guttman B, Ramanathan M, et al. Infections, vaccines and autoimmunity: A multiple sclerosis perspective. Vaccines (Basel). 2020;**8**(1):50. DOI: 10.3390/vaccines8010050. PMID: 32012815; PMCID: PMC7157658

[40] Bollard CM, Gottschalk S, Torrano V, et al. Sustained complete responses in patients with lymphoma receiving autologous cytotoxic T lymphocytes targeting Epstein-Barr virus latent membrane proteins. Journal of Clinical Oncology. 2014;**32**:798-808

[41] Straathof KCM, Bollard CM, Popat U, et al. Treatment of nasopharyngeal carcinoma with Epstein-Barr virus–specific T lymphocytes. Blood. 2005;**105**: 1898-1904

[42] Smith C, Tsang J, Beagley L, et al. Effective treatment of metastatic forms of Epstein-Barr virus–associated nasopharyngeal carcinoma with a novel adenovirus-based adoptive immunotherapy. Cancer Research. 2012;**72**:1116-1125

[43] Gottschalk S, Rooney CM. Adoptive T-cell immunotherapy. Current Topics in Microbiology and Immunology. 2015;**391**:427-454

[44] Münz C. Redirecting T Cells against Epstein-Barr virus infection and associated oncogenesis. Cell. 2020;**9**(6):1400. DOI: 10.3390/cells9061400. PMID: 32512847; PMCID: PMC7349826

[45] Pei Y, Wong JHY, Robertson ES. Targeted therapies for Epstein-Barr virus-associated lymphomas. Cancers (Basel). 2020;**12**(9):2565. DOI: 10.3390/cancers12092565. PMID: 32916819; PMCID: PMC7564798

[46] Slabik C, Kalbarczyk M, Danisch S, et al. CAR-T cells targeting Epstein-Barr virus gp350 validated in a humanized mouse model of EBV infection and lymphoproliferative disease. Molecular Therapy—Oncolytics. 2020;**18**:504-524. DOI: 10.1016/j.omto.2020.08.005. PMID: 32953984; PMCID: PMC7479496

[47] Hjalgrim H, Askling J, Rostgaard K, et al. Characteristics of Hodgkin's lymphoma after infectious mononucleosis. The New England

Journal of Medicine. 2003;**349**: 1324-1332

[48] Chen MR. Epstein–Barr virus, the immune system, and associated diseases. Virology. 2011;**2**:5

[49] Kimura H, Kwong YL. Epstein-Barr viral loads in diagnosis, monitoring, and response assessment. Frontiers in Oncology. 2019;**9**(62). DOI: 10.3389/fonc.2019.00062. PMID: 30809508; PMCID: PMC6379266

[50] Dierickx D, Habermann TM. Post-transplantation lympho proliferative disorders in adults. The New England Journal of Medicine. 2018;**378**(6):549-562

[51] Crombie JL, LaCasce AS. Epstein Barr virus associated B-Cell lymphomas and iatrogenic lymphoproliferative disorders. Frontiers in Oncology. 2019;**9**:109. DOI: 10.3389/fonc.2019.00109. PMID: 30899698; PMCID: PMC6416204

[52] Dharnidharka VR. Comprehensive review of post-organ transplant hematologic cancers. American Journal of Transplantation. 2018;**18**:537-549

[53] Parker A, Bowles K, Bradley JA, Emery V, Featherstone C, Gupte G, et al. Diagnosis of post-transplant lymphoproliferative disorder in solid organ transplant recipients - BCSH and BTS guidelines. British Journal of Haematology. 2010;**149**:675-692

[54] Gotoh K, Ito Y, Ohta R, et al. Immunologic and virologic analyses in pediatric liver transplant recipients with chronic high Epstein-Barr virus loads. The Journal of Infectious Diseases. 2010;**202**:461-469

[55] Lupo J, Germi R, Lancar R, et al. Epstein-Barr virus biomarkers have no prognostic value in HIV-related Hodgkin lymphoma in the modern combined antiretroviral therapy era. AIDS. 2019;**33**(6):993-1000

[56] Habib M, Buisson M, Lupo J, et al. Lytic EBV infection investigated by detection of soluble Epstein-Barr virus ZEBRA in the serum of patients with PTLD. Scientific Reports. 2017;**7**(1): 10479. DOI: 10.1038/s41598-017-09798-7. PMID: 28874674; PMCID: PMC 5585268

[57] Drouet E. The Role of the Epstein-Barr Virus Lytic Cycle in Tumor Progression: Consequences in Diagnosis and Therapy. 2019. DOI: 10.5772/intechopen.88607

[58] Feng W, Westphal E, Mauser A, et al. Use of adenovirusvectors expressing Epstein-Barr virus (EBV) immediate-early protein BZLF1 or BRLF1 to treat EBV-positive tumors. Journal of Virology. 2002;**76**:10951-10959

[59] Bilger A, Plowshay J, Ma S, et al. Leflunomide/teriflunomide inhibit Epstein-Barr virus (EBV)-induced lymphoproliferative disease and lytic viral replication. Oncotarget. 2017;**8**: 44266-44280

[60] Kawada JI, Ando S, Torii Y, et al. Antitumor effects of duvelisib onEpstein-Barr virus-associated lymphoma cells. Cancer Medicine. 2018;**7**:1275-1284

[61] Daskalogianni C, Pyndiah S, Apcher S, et al. Epstein-Barr virus-encoded EBNA1 and ZEBRA: Targets for therapeutic strategies against EBV-carrying cancers. The Journal of Pathology. 2015;**235**:334-341

[62] Germini D, Sall FB, Shmakova A, et al. Oncogenic Properties of the EBV ZEBRA Protein. Cancers (Basel). 2020;**12**(6):1479. DOI: 10.3390/cancers12061479. PMID: 32517128; PMCID: PMC7352903

[63] Balfour HH Jr. Progress, prospects, and problems in Epstein-Barr virus vaccine development. Current Opinion in Virology. 2014;**6**:1-5. DOI: 10.1016/j.

coviro.2014.02.005. Epub 2014 Mar 15. PMID: 24632197; PMCID: PMC4072744

[64] Sharma S, Rouce RH. Are we there yet? The never-ending quest for an Epstein-Barr virus vaccine. The Journal of Clinical Investigation. 2019;**129**(5): 1836-1838. DOI: 10.1172/JCI128370 PMID: 30985295; PMCID: PMC6486379